Model-Based Analysis and Optimisa
of Haber-Bosch Process Designs
for Power-to-Ammonia

Model-Based Analysis and Optimisation
of Haber-Bosch Process Designs
for Power-to-Ammonia

Von der Fakultät für Maschinenbau
der Technischen Universität Carolo-Wilhelmina zu Braunschweig

zur Erlangung der Würde

eines Doktor-Ingenieurs (Dr.-Ing.)

genehmigte Dissertation

von: M.Sc. Izzat Iqbal Cheema
aus (Geburtsort): Lahore

eingereicht am: 29.11.2018
mündliche Prüfung am: 21.02.2019

Gutachter:

Prof. Dr.-Ing. Ulrike Krewer
Prof. Dr.-Ing. Andreas Seidel-Morgenstern

2019

Bibliografische Information der Deutschen Nationalbibliothek

Die Deutsche Nationalbibliothek verzeichnet diese Publikation in der
Deutschen Nationalbibliografie; detaillierte bibliographische Daten sind im Internet
über http://dnb.d-nb.de abrufbar.

1. Aufl. - Göttingen: Cuvillier, 2019

   Zugl.: (TU) Braunschweig, Univ., Diss., 2019

© CUVILLIER VERLAG, Göttingen 2019

   Nonnenstieg 8, 37075 Göttingen

   Telefon: 0551-54724-0

   Telefax: 0551-54724-21

   www.cuvillier.de

1. Auflage, 2019

Gedruckt auf umweltfreundlichem, säurefreiem Papier aus nachhaltiger Forstwirtschaft.

   ISBN 978-3-7369-9995-4

   eISBN 978-3-7369-8995-5

Acknowledgements

First and foremost, I would like to thank my supervisor Dr.-Ing. Ulrike Krewer, for years of guidance, mentoring and technical insights and words of encouragement. Also thank you for providing me finances for participating in the conferences. I also thank Prof. Dr.-Ing. Andreas Seidel-Morgenstern for agreeing to be my second examiner and member of my Ph.D. Advisory Committee at the International Max Planck Research School for Advanced Methods in Process and Systems Engineering (IMPRS ProEng). I am grateful for your helpful discussions at various conferences and workshops. Furthermore, I thank Prof. Dr.-Ing. Stephan Scholl for the interest in my thesis, providing the helpful scientific discussions during various project meetings and finally for agreeing to chair my Ph.D. defence. Last but not least, I am also thankful to my colleagues, students, friends and family, here in Germany and back in Pakistan.

Finally, I am grateful to German Academic Exchange Service (DAAD) and Higher Education Commission (HEC), Pakistan for providing me scholarship for Ph.D. studies. Without their financial support, none of my achievements would have been possible.

Contents

List of Figures

List of Tables

Nomenclature

List of Symbols

A Constant for heat capacity / -

a Activity / -

A_{HE} Surface area for heat transfer / m^2

B Constant for heat capacity / -

C Constant for heat capacity / -

C^* Heat capacity rate ratio / -

C_p Specific heat capacity / kJ $kmol^{-1}$ K^{-1}

D Constant for heat capacity / -

E Constant for Henry's law coefficient / -

E_a Activation energy / kJ $kmol^{-1}$

F Constant for Henry's law coefficient / -

f Fugacity / -

G Constant for Henry's law coefficient / -

H Henry's law coefficient / bar

ΔH Heat of reaction / kJ $kmol^{-1}$

K Equilibrium constant / bar^{-2}

k Reaction rate constant / kmol m^{-3} h^{-1}

k_0 Frequency factor / kmol m^{-3} h^{-1}

K_c Vapour-liquid equilibrium constant / -

M Molecular weight / kg kmol^{-1}

$\dot{m}$ Mass flow rate / kg h^{-1}

$\dot{n}$ Molecular flow rate / kmol h^{-1}

NTU   Number of transfer units / -

P Pressure / bar

P^{sat} Vapour pressure / bar

Q Heat trasfer rate / kW

R Universal gas constant / 8.314 kJ kmol^{-1} K^{-1}

R_{NH_3} Rate of reaction / kmol m^{-3} h^{-1}

T Temperature / K

U Overall heat transfer coefficient / W m^{-2} K^{-1}

V Volume of catalyst bed / m^3

X Conversion of reactant / -

x Mass fraction / -

Y Mole percentage / mol %

y Mole fraction / -

z Manipulable process variables

Greek Symbols

α Constant / 0.5

β Vapour-feed molar fraction / -

γ Activity coefficient / -

ν Stoichiometric coefficient / -

ϕ Fugacity coefficient / -

ε Heat exchanger effectiveness / -

Superscripts

$*$ At a particular arbitrarily chosen standard state

o At temperature and pressure of system

Subscripts

2   Reverse reaction

b Catalyst bed

c Component

H   High

in   Inlet

L   Low

m Mixer

MAX   Maximum

MIN   Minimum

mix   Gas mixture

N   Normal

NOR   Normal

out   Outlet

$\textcircled{P}$ Gross product stream

q Quench stream

r Reactant

RS Reactor system

S Separator

$\textcircled{S}$ Stream

s   Shell side

t   Tube side

X   Extreme: highest or lowest

B   Bottom

SU   Separation unit

T   Top

Abstract

With an increase in global warming awareness, a clear shift of the world toward clean and sustainable processes can be seen. Specially, renewable energy-based electricity generation has gained the importance. With intermittent renewable energy in mind, chemical-based energy storage might be required and applied. Among several options, ammonia seems a promising carbon-neutral energy carrier. The power-to-ammonia concept allows for the production of ammonia from air, water and renewable energy. In principle, the subsequent ammonia synthesis loop is similar to the fossil fuel-based ammonia process, which has been developed over a period of one century. However, the operation and production flexibilities of ammonia synthesis reactor systems and loops have not yet been analysed and understood systematically, as the classical process is operated at its optimal steady state.

The objective of this dissertation is to analyse and understand the ammonia synthesis loop and autothermal reactor system with regard to flexible operation and production. Among several synthesis loop and autothermal reactor system possibilities, two synthesis loops and five autothermal three-bed ammonia synthesis reactor systems are considered. The synthesis loops vary in terms of ammonia separation unit allocations, $i.e.$ after and before a synthesis reactor system. The reactor system configurations differ in inter-stage cooling methods, which are based on either a direct cooling by quenching, a combination of indirect and direct cooling, direct and indirect cooling, or indirect cooling by heat exchange between process streams.

In the first part of this dissertation, the impact of the six process variables: operational pressure, process feed temperature, process feed composition (H_2-to-N_2 ratio, NH_3 and inert gas concentration) and feed flow rate on the flexible operation of an autothermal reactor system is quantified. Among them, H_2-to-N_2 ratio, inert gas concentration and feed flow rate provide high flexibilities in operation and production. Then, the effect of these process variables on flexibility is compared among five variants of autothermal reactor systems. All the reactor systems showed their feasibility for power-to-ammonia.

In the second part of this dissertation, multi-variable optimisation is applied for enhancing the load ranges of the reactor systems in the synthesis loop. Beside the reactor systems, the two synthesis loop configurations are also compared for hydrogen intake, ammonia production and recycle load flexibilities. With the multi-variable optimisation, the load range of all the synthesis loop and reactor system configurations drastically increases.

From this work it is concluded that all the ammonia synthesis reactor system and loop configurations allow operation for a large load range span and therefore are in principle suitable for power-to-ammonia.

Kurzfassung

Motiviert durch die Folgen der globalen Erwärmung kann ein deutlicher Umschwung in Richtung einer sauberen und nachhaltigen Energiegewinnung wahrgenommen werden. Hierbei rückt besonders Elektrizität aus z.b. Sonnen-, Wind- oder Wasserenergie in den Fokus. Aufgrund des fluktuierenden Charakters dieser erneuerbaren Energiequellen stellt die chemisch-basierende Energiespeicherung einen essentiellen Bestandteil der sogenannten Energiewände dar. Aus der Menge möglicher chemischer Energieträger sticht Ammoniak als vielversprechendster Kohlenstoffdioxid-neutraler Energieträger hervor. Das Konzept Power-to-Ammonia beschreibt eine umweltfreundliche Variante der Ammoniakproduktion aus Luft, Wasser und erneuerbarer Energien. Diese ähnelt grundsätzlich der klassischen Ammoniakproduktion aus fossilen Ressourcen, welche im Zuge des letzten Jahrhunderts ausgiebig erforscht und weiterentwickelt wurde. Da klassische Ammoniaksyntheseanlagen stets für eine optimale Ausbeute und somit für einen optimalen stationären Betriebspunkt ausgelegt sind, wurde bei diesen nicht auf einen dynamischen sowie flexiblen Betrieb eingegangen, weshalb sie nicht für Power-to-Ammonia eingesetzt werden können.

Das Ziel dieser Dissertation ist die Analyse und das Verständnis von autothermen Reaktorsystemen und deren Verhalten im Ammoniaksyntheseprozess mit dem Anspruch einer flexiblen Produktion. In einer Reihe von unterschiedlichen Konfigurationsmöglichkeiten werden zwei Synthesekreisläufe und fünf Reaktorsysteme mit jeweils drei Katalysatorbetten berücksichtigt. Das Prozesslayout unterscheidet sich hauptsächlich in der Lokalisierung der Ammoniakabscheidung, z.B. vor oder nach dem Reaktorsystem. Die Reaktorkonfigurationen variieren in der Art der Zwischenkühlung, welche durch direkte Einspritzung (Quenching), indirekt durch interne Wärmetauscher oder einer Kombination aus Quenching und indirekter Kühlung erfolgt, wobei es verschiedene Kombinationsmöglichkeiten von indirekter Kühlung und Quenching gibt.

Im ersten Abschnitt dieser Dissertation wird der Einfluss der sechs wichtigsten Prozessvariablen Betriebsdruck, Materialzufuhr-Temperatur, Materialzufuhr-Komposition (H_2-N_2-Verhältnis, NH_3- und Inertgaskonzentration) und Materialzufuhr-Massenstrom auf die mögliche Flexibilität des autothermen Reaktorsystems quantifiziert. Unter diesen Variablen stellen sich insbesondere das H_2-N_2-Verhältnis, die Inertgaskonzentration sowie der Massenstrom als effektive Stellgrößen heraus. Anschließend wird der Einfluss dieser Variablen auf die Prozessflexibilität der fünf untersuchten Prozesskonfigurationen untersucht und die jeweiligen Ergebnisse verglichen. Hierbei können alle Systeme die Anforderungen an einen effizienten und flexiblen Prozess erfüllen.

Im zweiten Abschnitt dieser Dissertation wird eine multidimensionale Optimierung durchgeführt, um das Betriebsfenster und die Produktionskapazität der Reaktorsysteme zu maximieren. Zusätzlich zu den Reaktorsystemen werden ebenfalls die Prozessvarianten in Bezug auf Wasserstoffverbrauch, Produktionskapazität und Umlaufmassenstrom verglichen. Mit Hilfe der multidimensionalen Optimierung konnte das Betriebsfenster aller untersuchten Systeme deutlich erweitert werden.

Diese Arbeit zusammenfassend kann gesagt werden, dass alle betrachteten Prozesskonfigurationen Produktionsflexibilität für ein weites Durchsatz-Betriebsfenster ermöglichen und somit für Power-to-Ammonia geeignet sind. Die Ergebnisse dieser Dissertation helfen autotherme Reaktorsysteme und Power-to-Amonia weiterzuentwickeln und somit eine flächendeckende Nutzung erneuerbarer Energiequellen zu ermöglichen.

1. Introduction[1]

In the twenty-first century, along with a rise in global warming awareness, zero-emission processes have gained substantial attention. Therefore, instead of fossil or nuclear energy fuelled power plants, the world is shifting toward renewable energy generation parks. A problem with renewable energy, however, is that it is seasonal, intermittent and decentralised when harvested. Thus, backup energy storage is required for uninterrupted and regulated power supply.

1.1. Ammonia as energy carrier

Chemicals-based storage is capable of high power and high capacity storage (see figure 1.1a) at low cost (see figure 1.1b) for longer seasonal time duration.[1,2] From figure 1.1a, chemical-based energy storage consisting of methane, methanol and ammonia seems promising with regard to large capacity storage over a longer period of time. Among these three options, ammonia is the only carbon free fuel. Ammonia is not just a carbon free fuel, it also provides higher energy densities than hydrogen, high round trip efficiency, scalability, availability of transport grid, maintained safety record and its production and consumption may be done CO_2-emission free.[3,4] As an energy carrier, ammonia has clear advantages over pure hydrogen, although hydrogen is considered to be a noble energy carrier, as it only produces water and energy. 1 mol of ammonia contains 1.5 mol of hydrogen, which is 17.8 wt % or 108 kg H_2/m^3 in liquid ammonia at 20 °C and 8.6 bar. Liquid ammonia contains 1.77 times more H_2 per unit volume than liquid hydrogen itself and even 4 times more than the most modern hydrogen storage methods in metal hydrides *i.e.* up to 25 kg H_2/m^3.[5]

Ammonia is the second most produced industrial chemical, and the production process has been intensively developed over a period of a century. Ammonia is used as a raw material for the production of various nitrogen compounds, including nitric acid, and a variety of fertilisers and polymers. Also, ammonia is used as refrigerant and neutraliser for NO_x emission from fuel combustion.[6] Moreover, ammonia has been tested and applied as fuel in compression ignition

---

[1]Part of this chapter has been published in I. I. Cheema and U. Krewer. *RSC Adv.*, 8:34926–34936, 2018.

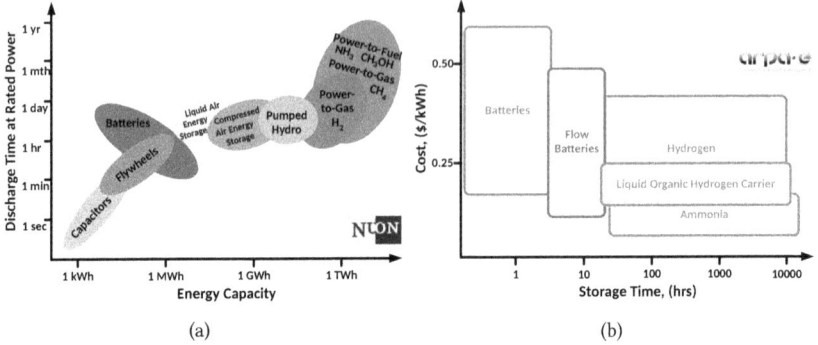

Figure 1.1.: Comparison between energy carriers for (a) capacity of energy storage[1] and (b) cost of energy storage.[2]

engines[7-9], spark ignition engines[10-12], gas turbines[13-15] and fuel cells[16,17] over a period of time. Despite its toxicity, ammonia has an excellent safety record in the fertiliser industry and a well established transportation network.[18,19] Thus, an ammonia economy would be low in cost and easier to apply than hydrogen in the energy sector.

Currently, about 1.6 % of fossil fuel, such as coal and natural gas, is used worldwide for the manufacturing of ammonia.[6] The classical production method, the Haber-Bosch process, relies heavily on natural gas[20], whereas ammonia has also the capability of being produced from renewable energy sources e.g. solar[21] and wind.[3,4,22]

1.2. Conventional ammonia production

versus power-to-ammonia

The conventional ammonia production process consists of three major sections: the steam and/or air based reforming of natural gas for producing H_2, CO_2 and the desired H_2-to-N_2 molar ratio, afterwards the removal of CO_2 and CH_4, and eventually the conversion of H_2 with N_2 to NH_3 in the Haber-Bosch reactor. In total, the conventional ammonia process consists of seven gas-solid catalytic steps: desulphuriser unit, primary reformer, secondary reformer, high temperature shift converter, low temperature shift converter, methanator and finally ammonia synthesis reactor.[6] An option for a CO_2 neutral production process is power-to-ammonia, which relies on H_2 production by splitting of water *via* electrolysis, where N_2 will be separated from air *e.g.* by pressure swing adsorption (PSA) and cryogenic distillation.[4] Energy requirements of pressure swing adsorption are up to 35 % less than the cryogenic distillation[23], therefore the air separation process based on pressure swing adsorption is considered in figure 1.2. The Haber-Bosch ammonia synthesis loop itself has shown to be similar to the conventional one.[4,21,22] The overview of the conventional *versus* the power-to-ammonia process is given in figure 1.2.

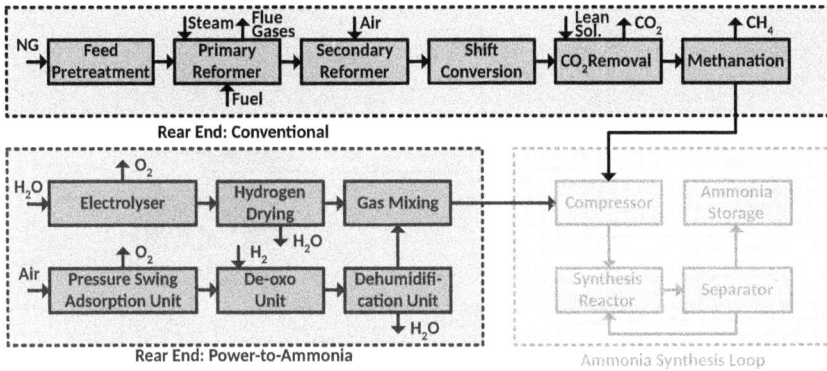

Figure 1.2.: Block diagram of the conventional ammonia production process and the power-to-ammonia process.

The efficiency of power-to-ammonia is estimated between 50 and 60 %, including hydrogen and nitrogen production[24], which is lower than from the latest classical Haber-Bosch ammonia production plants *i.e.* between 60 and 64 %.[25] This is mainly due to the high energy requirements and energy losses in the production of H_2 from electrolysis of water by atmospheric alkaline, high pressure alkaline (HPA - 16 bar) or proton exchange membrane electrolysers in comparison to energy requirements of the air separation based on cryogenic distillation or pressure swing adsorption and to the Haber-Bosch NH_3 synthesis process with iron, ruthenium or cobalt molybdenum bimetallic nitride based catalyst[24].

1.3. Power-to-ammonia-to-power

Fuhrmann *et al.*[4] reviewed the electro- and thermo-chemical sustainable ammonia based energy production and usage concepts illustrated in figure 1.3. Among these processes, the combination of a Haber-Bosch NH_3 synthesis loop with the H_2 supply from the electrolysis of water and the N_2 supply from the air separation *via* pressure swing adsorption or cryogenic distillation seems promising with respect to production capacities. They also discussed the potential for dynamic or flexible operation of the developed Haber-Bosch process concept, and as such, its ability to flexibly store excess renewable energy. With the growth of renewable energy production, power-to-ammonia and ammonia-to-power has gained world-wide interest. The current activities related to renewable ammonia in the U.S., Europe and Japan are comprehensively highlighted by Pfromm.[26]

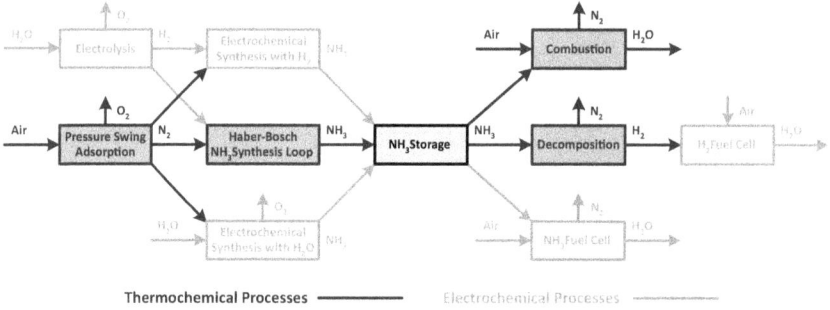

Figure 1.3.: Pathways of sustainable ammonia based energy production and usage concepts.

For the power-to-ammonia concept *via* the Haber-Bosch synthesis loop, a Technology Readiness Level of 6 has already been accomplished by Proton Ventures BV, The Netherlands.[21] The first pilot plant has been operating at West Central Research and Outreach Center, Morris, Minnesota, USA since 2013[22] and the second demonstrator became operational in June 2018 at Science & Technology Facilities Council's Rutherford Appleton Laboratory, Oxfordshire.[27] The operation of a power-to-ammonia plant by West Central Research and Outreach Center, Morris, Minnesota, USA has yet only been studied at steady-state, but not dynamically.

In parallel, simulations of the power-to-ammonia process were carried out for a system consisting of electrolyser, cryogenic separation and Haber-Bosch by Sánchez & Martín[28], whereas Cinti *et al.*[29] considered a process based on low temperature and high temperature electrolysers, pressure swing adsorption and Haber-Bosch. Cinti *et al.* analysed energy performances along with electricity consumption of every individual section. For the Haber-Bosch loop, thermodynamic equilibrium was considered instead of a kinetic approach, which is suitable for design-based analysis only. In contrast, Sánchez & Martín carried out complete system simulation and operational optimisation, including a kinetic approach for Haber-Bosch synthesis reactor. However, they did not consider autothermic operation of the ammonia synthesis reactor, which is of high interest for realising stand-alone power-to-ammonia plants. The question of how much an autothermal Haber-Bosch reactor system can be operated flexibly outside its standard conditions, is of crucial relevance for the power-to-ammonia concept, but has not been addressed so far. An alternative approach is to realise constant NH_3 production for the proposed power-to-ammonia process, mainly with help of an uninterrupted reactants supply. The uninterrupted supply of the reactants is maintained either by the continuous production of reactants with the help of non-stop supply of electricity or *via* producing excess amount of reactants during surplus energy which are stored and later on used during shortfall times.[1] However, storing H_2 reactant in bulk over a day can be up to three times more expensive than ammonia (figure 1.1b); in fact an ammonia storage tank is the cheapest way of storing large amounts of energy, *i.e.* greater than 100 GWh.[1,2]

As alternative to the above mentioned power-to-ammonia process, a low pressure reaction-absorption process for power-to-ammonia is suggested by Malmali *et al.*[30], whereas Wang *et al.*[31] proposed a power-to-power system design concept of *ca.* 72 % round trip efficiency. Malmali *et al.* used a synthesis loop consisting mainly of three steps: reaction, absorption and compression. They used an absorber instead of a condenser for separating NH_3 from unreacted gas, which allows one to operate the synthesis loop at 10 times lower pressure than the conventional process. This means lower capital and operational cost. However, work on absorbent regeneration still needs to be done.[30] Wang *et al.* used a H_2 powered cell in combination with a Haber-Bosch reactor. They divided their proposed ammonia-based energy storage system with respect to operation into two modes *i.e.* charging and discharging modes. During charging mode, heat is released from the NH_3 synthesis process and used in heating feed streams of reversible solid oxide fuel cell whereas waste heat contained by the product streams of reversible solid oxide fuel cell is used in reforming the H_2 during discharging mode.[31]

All the above mentioned processes mainly considered the Haber-Bosch design concepts for ammonia synthesis. However, the differences among the processes arise from the preparation of reactants, operational conditions of synthesis loop and/or ammonia separation. In this work, the focus is made on the developed Haber-Bosch process, rather than underdeveloped concepts.

1.4. Haber-Bosch ammonia synthesis loops

Ammonia synthesis is an exothermic reaction and is driven by thermodynamic equilibrium: for every 1 mole of N_2, 3 moles of H_2 are consumed to produce 2 moles of NH_3, see equation 1.1.

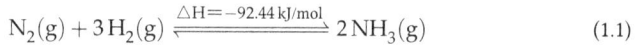

$$N_2(g) + 3H_2(g) \xrightleftharpoons{\triangle H = -92.44\,kJ/mol} 2NH_3(g) \tag{1.1}$$

Only partial conversion of the reactant feed (25 to 35 %) takes place in the synthesis reactor system. Therefore, reactants need to be separated from ammonia and recycled back. Ammonia is separated from un-reacted reactants by condensation, and preferably condensation takes place at low temperature and low temperature is achieved by passing through a trail of heat exchangers and coolers. A number of different ammonia synthesis loop configurations are possible, mainly these are classified with regard to the location of the ammonia separator and fresh feed entrance [6], see figure 1.4. Also, these ammonia synthesis loop configurations affect process feed composition and reactor design. [32]

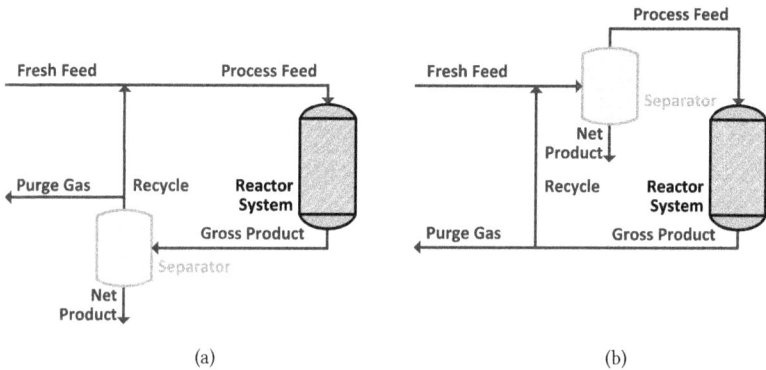

Figure 1.4.: Haber-Bosch ammonia synthesis loop configurations for (a) pure and dry fresh feed and (b) impure and wet fresh feed.

In power-to-ammonia, for a fresh feed two possibilities exist, pure or impure, and dry or wet. For example, for obtaining pure and dry reactants H_2 and N_2, post-treatment after electrolyser and pressure swing adsorption is required, see figure 1.2, otherwise in addition to argon (Ar), the feed may also contain H_2O and O_2. A pure and dry fresh feed is mixed with the recycle stream and is directly sent to the synthesis reactor system; afterwards it is passed through the ammonia separator, see figure 1.4a. For an impure and wet fresh feed, *e.g.* containing catalyst poisons like CO_2, O_2 and H_2O, fresh feed is mixed with the recycle stream before transporting it to the synthesis reactor system and then passing it through the NH_3 separator. This way, impurities can be completely dissolved by condensing NH_3; see figure 1.4b. Both loops require purging; purging of gas takes place for maintaining a low inert gas concentration within the synthesis loop. For power-to-ammonia, argon (Ar) is a possible inert gas which is transported along with N_2. With regard to minimum energy requirements, maximum NH_3 separation and minimum NH_3 concentration in the recycle stream, the best synthesis loop configuration is shown in figure 1.4a. [6,32]

Depending upon the storage capacity, there are two different methods used for liquid ammonia storage: for small quantities *i.e.* below 1500 t, ammonia is stored in spherical stainless steel tanks under *ca.* 10 bar pressure at ambient temperature and for large quantities up to 50 000 t, ammonia is stored in liquid form at about 1 bar and $-33\,^{\circ}\text{C}$[6]. Power-to-ammonia requires decentralised production units with small quantity of ammonia storage, as renewable energy harvesting is decentralised. As such, liquefaction of ammonia for energy storage from renewables is preferably done at atmospheric temperature by applying the required pressure.

In addition to being highly energy efficient and allowing lower storage capacities, power-to-ammonia requires flexibility in operation and ammonia production. Therefore, it would be of interest to inquire into the suitable synthesis loop configuration for a flexible ammonia production. During operation of the power-to-ammonia pilot plant at Morris, Minnesota, USA, it was determined that the NH_3 production rate in the Haber-Bosch synthesis loop is mainly influenced by the reactor system, the NH_3 separation by condensation and the recycle loop. The reactor system is a bottleneck of the synthesis loop, as the reactor system applies the strongest influence on the NH_3 production: at least three times higher than the other two.[22] Therefore, inquiring into the suitable synthesis reactor system configuration for flexible ammonia production will be of great importance.

1.5. Ammonia synthesis reactor systems

Commercially available ammonia synthesis reactor systems are mainly classified into two main types: tube-cooled and multi-bed. For tube-cooled reactor systems two configurations are possible, *i.e.* cooling tubes are fixed within the catalyst bed or catalyst inside the tubes and cooling medium on the shell side. Mostly, the cooling medium is process feed gas. For multi-bed reactor systems, the catalyst is spread over various beds with different interstage heat exchanger configurations: direct quenching of reactant gas with product gas (*e.g.* Chemico and Pullmann-Kellogg Converters), quenching of reaction mixture with help of external cooling (*e.g.* Montecatini Edison Converters) and exchange of heat with reacting gas (*e.g.* Tennessee Valley Authority (TVA) Converters). A comparison of ammonia synthesis reactor systems is given by Strelzoff.[33] In ammonia synthesis reactor systems, the temperature conditions for inlet and outlet are preferably managed by exchanging heat between process streams. The heat of reaction (equation 1.1) is itself sufficient for maintaining the temperature level in the reactor system, which allows one to operate the process autothermically. In recent work, tube cooled and multi-bed reactor systems are compared.[34,35] Overall, the tube cooled reactor systems are more efficient in recovering the heat of reaction. However, costs for the tube cooled reactor system along with the catalyst are greater than three times the cost of the quench cooling reactor system.[35] In addition, at optimum conditions, tube cooled and indirect cooling reactant conversion profiles are mostly identical and resulted in the same overall N_2 conversion.[34] In multi-bed reactor systems, Khademi & Sabbaghi[34] considered two types: quench and indirect cooling, with combinations of two, three and four catalyst beds. Among these three types of reactor systems, the optimisation results for maximum N_2 conversion are compared at the same conditions: catalyst volume, reactor pressure and feed mass flow rate. Among these reactor systems, the three-bed reactor system is the most efficient for NH_3 production, energy savings, capital and maintenance cost. The indirect cooling reactor system results in higher N_2 conversion from the quench type. Therefore, three-bed autothermal reactor systems are considered in this work.

1.6. Scope of this work

Over a period of a century, conventional ammonia production processes have been developed for large capacities and preferably operated at steady-state conditions only, as H_2 reactant is economically producible by reforming of fossil fuels. However, power-to-ammonia relies on H_2 production from electrolysis of water with high consumption of intermittent renewable energy, and as a consequence, requires decentralised and flexible operation. The Haber-Bosch ammonia synthesis process physically remains the same for both processes, although the requirements change. The ammonia synthesis process consists of complex interconnected systems, such as mixing and compression units, synthesis reactor system, a trail of heat exchangers and coolers, separation unit, recycle loop and storage tank. For power-to-ammonia, the ammonia synthesis loop's flexible operation and NH_3 production are not yet well understood. Therefore, the research hypothesis of this dissertation can be described as follows:

"Flexible ammonia production through the Haber-Bosch process is possible by selecting a suitable reactor system design and loop configuration, and with implementation of multi-variable optimisation"

The objectives of this dissertation are as follows:

1. Configure an ammonia synthesis reactor system for autothermal operation.

2. Investigate the operating envelope of the autothermal reactor systems through steady-state stability analysis.

3. Enhance the load range of the Haber-Bosch process designs by applying multi-variable optimisation.

4. Perform a flexiblity analysis for the ammonia synthesis loops.

In this work, model-based analysis is performed *via* investigating the most suitable Haber-Bosch process designs among the studied design variants for renewable energy based operation. With regard to design variations, the reactor system and synthesis loop designs are altered. For the reactor system, a three-bed autothermal configuration is considered and alterations are carried out in terms of interstage cooling types, *e.g.* quench-based (direct), heat exchanger-based (indirect), combination of quench and heat exchanger-based. For the synthesis loop configurations, ammonia separation after and before the reactor system are considered. In chapters 2 and 3, ideal ammonia separation is considered and ammonia is recycled back in small quantity for maintaining the desired process feed composition. In chapters 4 and 5, an ammonia separation unit model is included. The outline of this dissertation is as follows:

In chapter 2, a three-bed autothermal quench based inter-stage cooling reactor system is considered. The emphasis is given on investigating the operating envelope of the reactor system by varying one process variable at a time *e.g.* operation pressure, process feed temperature and composition (reactants ratio, NH_3 and Ar concentrations). Afterwards, effect of the process variables on NH_3 production, H_2 intake, recycle load and recycle to fresh feed ratio are investigated. In chapter 3, three-bed autothermal reactor systems with five different inter-stage cooling types, *e.g.* quench-based, heat exchanger-based and combination of both types are considered for minimum and maximum ammonia production. The effect of process variables, like the inert gas percentage in the synthesis loop, the H_2-to-N_2 ratio and the total feed flow rate on ammonia production for the design variants are investigated.

In chapter 4, two synthesis loop configurations with three-bed autothermal quench based inter-stage cooling reactor system are considered. The synthesis loops vary with regard to ammonia separation unit allocation, *i.e.* ammonia separation after and before the reactor system. For both loops, volumes of the reactor systems catalyst beds are optimised while keeping the same synthesis loop conditions, such as, process feed composition, operation pressure and process feed temperature. Furthermore, the comparison between synthesis loops is made with regard to minimum and maximum ammonia production by the implementation of multi-variable optimisation. In chapter 5, the concepts of chapters 3 and 4 are combined to investigate the load range of the Haber-Bosch process designs. Here, the inclusion of an ammonia separation unit after the reactor system configuration is considered for the synthesis loop for five different reactor system designs. Multi-variable optimisation is applied with guaranteed autothermal operation in the reactor systems. The comparison between design variants is also made for NH_3 production, H_2 intake, recycle load, recycle to fresh feed ratio and purge ratio.

Finally, the findings of this work are summarised along with future aspects and shortcomings in chapter 6.

2. Operating envelope of Haber-Bosch process design[1]

In this chapter, the operating envelope for steady state operation of a three bed autothermic Haber-Bosch reactor system is determined by means of systemic and stability analysis.

2.1. Haber-Bosch synthesis loop and reactor system

The Haber-Bosch ammonia synthesis loop for producing NH_3 consists of mixing and compression units, synthesis reactor system, a trail of heat exchangers and coolers, a separator, a recycle loop and a storage unit. Altogether, it can be divided into four subsections, as shown in figure 2.1. The system design of the ammonia synthesis reactor poses a challenge due to the harsh reactor requirements of high inlet temperature to achieve high reaction rate and simultaneously, low outlet temperature to achieve a high equilibrium conversion.[36] Furthermore, a high reactant conversion should be achieved despite constraints due to equilibrium conversion. This is accomplished through the use of several catalyst beds in series.[37] The usual operational envelope ranges are: pressure of 150 to 300 bar, temperature of 623 to 773 K, H_2-to-N_2 molar ratios of 2 : 1 to 3 : 1 and inert gas content from 0 to 15 mol %.[6] The operational envelopes mentioned above for carrying out the ammonia synthesis reaction are quite general, and vary greatly. However, Haber-Bosch process plants have some constraints imposed due to design[32,34] and operation limitations[38], which originate from requirements of autothermic operation of the reactor system, catalyst type, feed content and composition. Therefore, the operating envelope needs to be determined and customised with respect to the process plant. Furthermore, due to low conversion (25 to 35 %), un-reacted reactants need to be separated and recycled back.[6] Therefore, the recycled reactants flow rate (recycle load) is several times higher than the feed flow rate. In the power-to-ammonia synthesis loop, the only inert gas is argon[24], originating from the air separation unit, along with the N_2 used as a reactant. In the conventional process, inert gases are CH_4 and Ar.[6] Concentration of Ar in the synthesis loop is controlled by purging a small amount of gas from the recycle stream.[24]

[1]Part of this chapter has been published in I. I. Cheema and U. Krewer. *RSC Adv.*, 8:34926–34936, 2018.

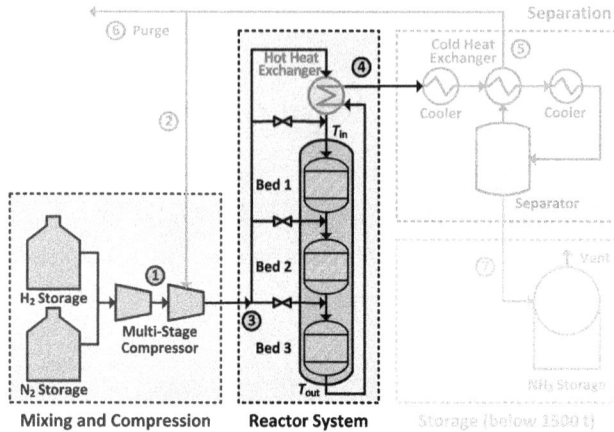

Figure 2.1.: Ammonia synthesis loop with small quantity ammonia storage for power-to-ammonia.

During the power-to-ammonia pilot plant operation at Morris, Minnesota, USA it was determined that the production of ammonia is controlled by three bottlenecks in the ammonia synthesis loop: catalytic reaction, NH_3 separation by condensation and recycling of un-reacted
reactants. Among these production bottlenecks, catalytic reaction has at least three times
higher influence than the others.[22] In an ammonia synthesis reactor system, the temperature conditions for inlet and outlet are managed by exchanging heat between outlet and inlet
streams. The heat of reaction is itself sufficient for maintaining the temperature level in the
reactor system, allowing the process to be operated autothermically, see figure 2.1. However,
this requires careful heat management in the reactor system, particularly between inlet and
outlet streams. If the inlet stream is not sufficiently heated, the rate of reaction will drop and
will lead to lower outlet temperature, which results in lowering inlet temperature and eventually the reaction will stop completely.[39] Therefore, the analysis and careful operation of the
ammonia synthesis reactor system in an ammonia synthesis loop carries great importance for
design and off-design operation.

Much of the work regarding the ammonia synthesis reactor system revolved around an in-
cident that occurred in an industrial ammonia fixed-bed synthesis reactor in Germany in
1989.[40] Multiplicity of periodic behaviour and stability analysis of ammonia reactor systems
are repeatedly mentioned in the literature.[40–44] But much of the work only highlighted the

effect of reactor operational pressure, inlet temperature and feed temperature, and did not consider feed flow rate and feed composition e.g. H_2-to-N_2 molar ratio, NH_3 and inert gas concentration. These variables, though, would be essential to operate and control a power-to-ammonia system flexibility. Morud and Skogestad in 1998 analysed the Haber-Bosch process with a pseudo-homogeneous dynamic model for a three catalyst bed reactor system and a static model for a counter current heat exchanger[40], Mancusi et al. in 2000, 2001 and 2009 analysed the same process with a heterogeneous model and concluded substantial qualitative agreement with the pseudo-homogeneous results e.g. shutdown pressure and feed temperature for the reactor system was more than the pseudo-homogeneous by about 18.57 bar[42] and 20 K.[43] Azarhoosh et al.[45] also considered a one-dimensional heterogeneous model, and compared results with the real plant and had differences of up to 13.5 K in the catalyst beds. In addition, they also optimised the synthesis reactor for maximum ammonia production by adjusting input temperature, total feed flow rate and operating pressure. Farivar & Ebrahim[46] extended this work by using a two-dimensional model and a finite volume method. In comparison to their previous work[45] they reduced the temperature difference to 4 K in the catalyst bed from real plant data. They also analysed the effect of pressure. Furthermore, a simple dynamic model-based stability analysis for a single bed ammonia synthesis reactor and heat exchanger was studied by Rabchuk et al.[44] for a step change of the parameters of pressure, temperature and flow rate. They concluded that a more realistic thermodynamic model needs to be added, and that the reactor system should be extended to a higher number of catalyst beds, corresponding to the real ammonia synthesis reactor system. Among multi-bed reactor systems, e.g. two to four catalyst beds, the three bed reactor system is the most efficient and cost effective for NH_3 production.[34] The operational and production flexibility for the conventional ammonia synthesis reactor system has not yet been systematically analysed, as the plants are mostly designed for large capacities and the raw material methane is abundantly available and easily storable at highly constant inlet conditions.

The focus of this chapter is to determine the steady-state operational and production limitations of the ammonia synthesis reactor system and recycle loop, as renewable energy will be only intermittently available for the production of the reactants. H_2 is the limiting reactant in the power-to-ammonia process, as more than 90 % of the energy is consumed during its production. During energy shortage periods, H_2 production may need to be reduced or even shut down.[24] Thus, knowing the operational flexibilities of the process variables, H_2 intake and NH_3 production flexibilities along with the change in recycle load and recycle to feed ratio is of high relevance and should be analysed. Therefore, the focus on such an analysis is necessarily needed. The quench based inter-stage cooling three bed ammonia synthesis reactor system (figure 2.1) with autothermal operation, *i.e.* energy sufficiency without additional heating/cooling is considered. Therefore, the pseudo-homogeneous mathematical model along with the assumptions of the reactor system are defined first. Afterwards, the effect of the following process variables is analysed: reactor pressure, inert percentage in synthesis loop, NH_3 concentration, H_2-to-N_2 ratio, total flow rate and inlet temperature of reactor system on the operational envelope, H_2 intake and NH_3 production flexibilities, along with change in recycle load and recycle to feed ratio for the reactor system.

2.2. Mathematical model and simulation

Physicochemical modelling is applied to analyse the ammonia synthesis reactor system under steady-state operation. The systematically applied approach subdivides the reactor system into three subsystems *i.e.* heat exchanger, catalyst beds and mixers. The processes taking place within the boundaries of each subsystem are distinguishable physically and/or chemically. By combining the individual subsystems, the behaviour of the overall synthesis system can be quantified. First, the simplifying assumptions, along with mathematical models, are presented. These models are then followed by simulation scenarios for identifying operation, H_2 intake and NH_3 production flexibilities for the reactor system along with the change in recycle load and recycle to feed ratio. To focus on the complex reactor system, the design and operational limitation which may originate from the separation section by the heat exchanger, coolers and the NH_3 separator to recycle stream has been ignored. Therefore, changes in recycle and recycle to feed ratio are independent of any kind of limitations. The detailed design and construction specifications of the reactor system are not within the scope of this chapter. Therefore, a pseudo-homogeneous reactor model is adapted and heat losses are ignored, though with this assumption, behaviour of the reactor system remains quite similar to a real plant.[42,43,45,47] Future studies may tailor the separation section to the required flexibility envelope of the Haber-Bosch process.

2.2.1. Subsystems models

In the following, the assumptions and physical equations for the subsystems are given.

Heat exchanger

All the fluids in the heat exchanger remain in the gaseous phase and as such no condensation is considered for modelling. Hot gas flows through tube side and cold gas flows through shell side of the heat exchangers.[40–44] The heat exchange between tube and shell side gas takes place using a combination of counter current and cross flow. The temperature of the gases changes in the axial direction of flow and does not change in its radial direction. Heat of conduction in the axial direction is also negligible.[48] All thermal properties of the gases and the exchanger wall are constant. No heat losses occur to the surroundings due to external insulation, *i.e.* the component is adiabatic. Chemical reaction and mass transfer do not take place. Therefore, the system can be described by a steady state energy balance and the feed-effluent heat exchanger is modelled by an ε-NTU model[40] using the effectiveness ε as follows:

$$T_{s_{out}} = \varepsilon \, T_{t_{in}} + (1 - \varepsilon) \, T_{s_{in}} \qquad (2.1)$$

where $T_{s_{out}}$ is the shell side outlet temperature and $T_{t_{in}}$ is the tube side inlet (catalyst bed 3 outlet) temperature, $T_{s_{in}}$ is the shell side inlet temperature, and ε is the heat exchanger effectiveness. The ε is constant, independent of change in inlet temperature and generally lies within the range 0.4 to 0.8 depending on the configuration of heat exchanger. In context to figure 2.1, the streams of the heat exchanger will be $T_{s_{out}} = T_{in}$, $T_{t_{in}} = T_{out}$ and $T_{s_{in}} = T_{③}$. The ε-NTU model has the advantage over conventional methods as it does not require evaluation of mean temperature differences and detailed design of the heat exchanger. The ε-NTU model is also suitable for solving off-design heat exchanger problems.[49] The thermal effectiveness (equation A.1) for shell and tube heat exchanger, along with specifications (see table B.1) are given in appendices A and B, respectively.

Catalyst bed

The heart of an ammonia synthesis reactor is the isobaric and adiabatic catalyst bed. The reaction takes place at the surface of the catalyst, where nitrogen and hydrogen are consumed, and ammonia is formed in an exothermic reaction. With consideration of a radial flow catalyst bed, a gradient of temperature and concentration (or partial pressure) is generated in radial direction, which also permit the handling of small diameter catalyst particles[6] with high catalyst efficiency[50] and almost negligible pressure drop[51], so isobaric conditions are assumed. For fine catalyst particles of size 1.5 to 3 mm, the rate of formation for ammonia can be taken without correction factors such as effectiveness factor and with consideration only for convective driving forces for transport of mass and heat between the flowing gases and catalyst.[50] Further, the temperature gradient ΔT inside the catalyst pellet is negligible, as high thermal conductivity magnetite Fe_3O_4 catalyst[52] is assumed. Therefore, heat transfer resistance between pellet and gas is also neglected. The steady state material and energy balance for the fine catalyst particles in catalyst beds are shown in equations 2.2 and 2.3, respectively:

$$\frac{dX_{r,\,b}}{dV_b} = \frac{\nu_r\,R_{NH_3,\,b}}{2\,\dot{n}_{r,\,b_{in}}} \tag{2.2}$$

$$\frac{dT_b}{dV_b} = \frac{(-\Delta H_b)\,R_{NH_3,\,b}}{\dot{m}_{b_{in}}\,C_{p_b}} \tag{2.3}$$

where subscript $r \in \{N_2 \text{ or } H_2\}$ refers to reactants and $b \in \{1, 2, 3\}$ to the three catalyst beds. ν is the stoichiometric coefficient, X is fractional conversion of reactant, V is the volume of the catalyst bed, R_{NH_3} is the reaction rate, $\dot{n}$ is the initial molar flow rate of reactant, T temperature of reacting mixture, ΔH is the heat of reaction, C_p is the specific heat of reacting mixture and $\dot{m}$ is the total mass flow rate of the reacting mixture. The conversion differential equations for both reactants are considered, instead of just one reactant, as during the analysis of the operational envelope for H_2-to-N_2 ratio, limiting reactant shifts between N_2 and H_2, which also requires one to change the differential equation. By using reactant conversion, the molar fractions of components are calculated by using equations A.4 to A.7, see appendix A.

The rate of reaction is calculated by a modified form of the Temkin equation[53], developed in 1968 by Dyson & Simon.[50] The activities are considered instead of partial pressures, as follows:

$$R_{NH_3} = k_2 \left(K^2 a_{N_2} \left(\frac{a_{H_2}^3}{a_{NH_3}^2} \right)^\alpha - \left(\frac{a_{NH_3}^2}{a_{H_2}^3} \right)^{1-\alpha} \right) \tag{2.4}$$

where a_{N_2}, a_{H_2}, a_{NH_3}, k_2, K and α are activity coefficients for nitrogen, hydrogen and ammonia (equations A.8 to A.11, appendix A), constant for reverse reaction (equation A.12, appendix A), equilibrium constant of reaction (equation A.13, appendix A) and constant (table A.1, appendix A), respectively. Also, the equations used for calculating specific heat C_p (equations A.15 to A.17) and heat of reaction ΔH (equation A.18) are stated in the appendix A.

Mixer

The mixing of gases in the mixer is assumed to be ideal and instantaneous. The heat of mixing is neglected, as components do not interact strongly with each other.[54] Also, pressure remains constant, as isobaric conditions are assumed in overall reactor system. The steady state material and energy balance for the adiabatic mixer are used as follows for calculating the reactant conversion and temperature after quenching:

$$X_{r,\,m_{\text{out}}} = \frac{\sum_{b=1}^{m-1}\left(\dot{n}_{r,\,b_{\text{in}}} \cdot \prod_{b=1}^{m-1} X_{r,\,b_{\text{out}}}\right)}{\sum_{b=1}^{m-1}\left(\dot{n}_{r,\,b_{\text{in}}} \cdot \prod_{b=1}^{m-1} X_{r,\,b_{\text{out}}}\right) + \left(\dot{n}_{r,\,b_{\text{out}}} + \dot{n}_{r,\,q_{\text{in}}}\right)} \tag{2.5}$$

$$T_{m_{\text{out}}} = \frac{\dot{m}_{b_{\text{out}}} C_{p_{b_{\text{out}}}} T_{b_{\text{out}}} + \dot{m}_{q_{\text{in}}} C_{p_{q_{\text{in}}}} T_{q_{\text{in}}}}{\left(\dot{m}_{b_{\text{out}}} + \dot{m}_{q_{\text{in}}}\right) C_{p_{m_{\text{out}}}}} \tag{2.6}$$

The mixers between the catalyst beds are considered in operation only *i.e.* mixer 2 and 3. Therefore subscript $q \in \{2,3\}$ refers to quench stream, $m \in \{2,3\}$ refers to mixers and $b \in \{1,2\}$ refers to beds.

Flexibility

The equations A.19 to A.35 used for calculating the material balance of streams ① to ⑦ mentioned in figure 2.1 for the ammonia synthesis loop are given in appendix A. The process variables operational flexibility, the H_2 intake and the NH_3 production flexibility are defined as a fractional change from the normal values:

$$\text{Flexibility} = \frac{\text{Actual} - \text{Normal}}{\text{Normal}} \times 100 \tag{2.7}$$

2.2.2. Simulation

The simulation is performed in MATLAB software and a built-in ODE solver (ode45) is used for the implementation of differential equations. For normal operation, the fresh stream ① N_2

supply with $2\,\text{mol}\,\%$ of Ar and pure H_2 supply from storage is considered in ratio of $3\,\text{mol}$ of H_2 to $1\,\text{mol}$ of N_2. Also, the fresh supply is considered free of impurities like H_2O and O_2. After the reactor system, unused reactants are separated from NH_3 and recycled back with the assumption that $27.79\,\text{mol}\,\%$ of NH_3 is carried along with them during normal operation. A concentration of $5\,\text{mol}\,\%$ of inert gas is maintained in the reactor system intake stream ③ by purging 0.0241 weight fraction of recycle stream ⑤. The initial conditions used are given in table 2.1, unless specified separately. The stream numbers are labelled in figure 2.1.

Table 2.1.: Initial conditions

Normal (N) operation streams composition				
Stream Number	$Y_{H_{2_N}}$ / mol %	$Y_{N_{2_N}}$ / mol %	$Y_{NH_{3_N}}$ / mol %	Y_{Ar_N} / mol %
①	74.62	24.88	0.00	0.50
③	68.12	22.71	4.17	5.00
Inlet & normal (N) operational conditions at reactor system			
$X_{r③}$ / -	$T_{③N}$ / K	P_N / bar	
0.00	523.00	200.00	

The catalyst bed volumes, feed flow rate and quench flow rates $\dot{m}_{q_1}$, $\dot{m}_{q_2}$ and $\dot{m}_{q_3}$ for the given normal operation and feed composition are adjusted by trial and error method for producing $120\,\text{kg/h}\,NH_3$, excluding the $1.11\,\text{kg/h}\,NH_3$ lost in purge gas. For achieving the optimal reactor design volume with the maximum possible reaction rate, inlet temperatures of all catalyst beds are maintained at $673\,\text{K}$ and their outlet temperature at $773\,\text{K}$ or $90\,\%$ of the equilibrium temperature. The reactor operation pressure is considered $200\,\text{bar}$ which is within the usual operational range mentioned earlier in first paragraph of chapter. With this NH_3 production capacity, an ammonia-to-power plant is capable of generating $50\,\text{MWh/day}$ of energy from ca. $3\,\text{tons/day}$ of ammonia via IC engine of $29\,\%$ efficiency.[21] For design only, the reaction is considered to be accomplished when reaching $90\,\%$ of the equilibrium composition, as for equilibrium conversion operation an infinite amount of reactor space is required.[37] Also, the reactants and the product present in purge stream were assumed to be lost. However, in reality reactants and ammonia are recovered from the purge stream through separation processes; most importantly, up to $95\,\%$ hydrogen is recoverable.[6] The breakdown of the reactor system for each catalyst bed volume and feed flow rate is shown in table 2.2.

The steady state operating envelope and stability for the autothermic reactor system is investigated with the help of van Heerden plot[39] for six process variables: reactor pressure, inert concentration, ammonia concentration, H_2-to-N_2 ratio, total flow rate and temperature at inlet

Table 2.2.: Catalyst bed volumes and normal operation flow rates

Reactor System	Bed 1	Bed 2	Bed 3	Total
V / m^3	0.0075	0.0221	0.0464	0.0760
$\dot{m}_{q_N}$ / kg h^{-1}	-	163.83	177.01	340.84
$\dot{m}_{b_N}$ / kg h^{-1}	321.70	485.53	662.54	662.54

stream ③ of the reactor system. During the steady-state stability analysis one process variable is changed and the other five process variables are held constant. The plots consist of two different kinds of graphs: the S-shaped heat production curve and the straight-line for heat removal, e.g. see figure 2.4. The S-shaped curve shows the relation between temperature of the reactor system bed 1 inlet (T_{in}) and bed 3 outlet (T_{out}), rise in temperature is due to exothermic reaction, the straight-line shows the characteristics of heat exchange in the heat exchanger (HE). With help of the heat exchanger, heat is transferred from the bed 3 outlet stream to the bed 1 inlet stream; at steady state operating points, both lines intersect. Under many given operating conditions, multiple steady-states, i.e. intersection point of heat production and heat removal lines are obtained. As such, the reactor system can work up to three different steady states characterised by the different temperatures of bed 1 and 3. The lower steady state point and upper steady state point are stable, the upper steady state point is desired for operation due to stability and maximum conversion. The middle steady state point will be unstable: with a minor increase in temperature, the heat of production rises more rapidly than the heat of removal and the temperature will continue increasing until the new point of intersection between heat of production and removal lines is met. For a minor decrease in temperature, the heat of production will continue declining until the point of intersection between heat of production and removal lines met.

2.3. Results and discussions

The results obtained from the model are presented and discussed in this section. First, the reactants fractional conversion and temperature profile along the reactor beds are presented for normal operation. Afterwards, stability analysis is performed for the six process variables to determine operational, H$_2$ intake and NH$_3$ production flexibilities along with change in recycle and recycle to feed ratio. See table 2.3 for results summary. The normal and boundary operation results for each bed inlet and outlet are summarised in table B.2, see appendix B.

2.3.1. Normal operation

Reactants conversion and temperature progression along the catalyst beds are shown in figures 2.2a and 2.2b, respectively. The hydrogen and nitrogen conversion profiles overlap, as the
reactants' ratio, H_2-to-N_2, is stoichiometrically balanced as 3 : 1 (see equation 1.1). Ammonia
synthesis is an exothermic reaction that releases heat and therefore the temperature along
each bed increases. The rise in reactants conversion and temperature occurs at much higher
rate in bed 1 than beds 2 and 3 due to low ammonia content and feed flow rate in bed 1. For
accommodating the higher ammonia content and feed flow rate, bed 2 and bed 3 are of larger
volume compared to bed 1.

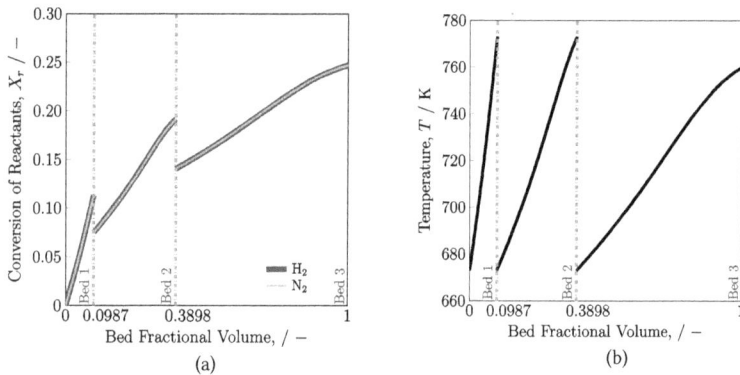

Figure 2.2.: For the reactor system: (a) reactants conversion profiles and (b) temperature profile along
the catalyst beds.

Reactants conversion *versus* temperature and the equilibrium line for the reactor system is
presented in figure 2.3. The solid lines represent temperature and reactants conversion within
catalyst beds, whereas dash dotted lines represent temperature and reactant conversion within
mixers. The reactor system is operated for the maximum possible reactants conversion and
temperature span. For the catalyst bed 3 the $\overline{TX}$ trajectory touches the operational (OP) line
i.e. 90 % of the equilibrium (EQ) line and reaction is stopped at 760 K as reactor volume was
chosen such that 90 % conversion may occur to avoid infinite amount of reactor space required
for reaching to equilibrium. The effectiveness of heat exchanger $\varepsilon = 0.6329$, which is calculated
by using equation 2.1 for normal operation temperature range. It remains constant during
stability analysis of the reactor system and help in determining the intersection temperature.

The reactants conversion and temperature from 773 to 673 K within mixers decrease due to quenching of fresh feed. Results summary for normal operation are presented in table B.2, see appendix B.

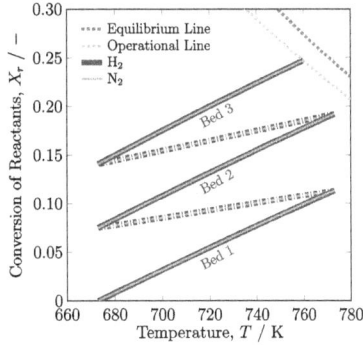

Figure 2.3.: Temperature-reactants conversion $\overline{TX}$ trajectories along the catalyst beds of the reactor system.

2.3.2. Operational and production flexibilities

In the following subsection, the operating envelope, *i.e.* the lower (L) and higher (H) operating points of the autothermic reactor system for the main process variables: reactor pressure, inert concentration, ammonia concentration, H_2-to-N_2 ratio, total flow rate and temperature at the inlet of the reactor system is analysed. The summary of operating envelope, operational flexibility of the respective process variable, hydrogen intake and ammonia production flexibilities, along with the resulting change in recycle load and recycle to feed ratio is given in table 2.3.

The stability analysis for the reactor pressure is presented in figure 2.4. For the normal (N) reactor operation at 200 bar, it is required that the feed must enter bed 1 at 673 K. For lower temperatures, the reactor will not be able to produce sufficient heat to maintain the reaction, and the inlet temperature at bed 1 would move towards unstable steady state temperature *ca.* 644 K. Further cooling from this point will result in the shut down of reactor system, due to more heat removal than heat production. Likewise, the heat production curve can be moved up and down by changing reactor pressure, until it intersects the heat removal curve at two or one point(s) instead of three points *i.e.* from 194.32 to 235.76 bar or onwards. The increase in pressure increases reactants conversion (see table B.2, appendix B) due to higher reaction

Table 2.3.: Reactor system operating envelope and operational flexibility of the process variables, as well as, $\dot{m}_{H_2 \text{①}}$ resulting H_2 intake and $\dot{m}_{\text{⑦}}$ NH_3 production flexibilities along with change in $\dot{m}_{\text{②}}$ recycle load and recycle to feed ratio $(\dot{m}_{\text{②}}/\dot{m}_{\text{①}})$

Process Variables	Operating Envelope		Flexibility			Change in	
		Operational	$\dot{m}_{H_2 \text{①}}$	$\dot{m}_{\text{⑦}}$	$\dot{m}_{\text{②}}$	$\dfrac{\dot{m}_{\text{②}}}{\dot{m}_{\text{①}}}$
		/ %	/ %	/ %	/ %	/ %
P / bar	194.32	Low	-2.84	-9.92	-10.14	$+2.49$	$+13.79$
	213.91	High	$+6.95$	$+5.57$	$+5.72$	-1.40	-6.60
$Y_{Ar\,\text{③}}$ / mol %	0.00	Low	-100.00	$+15.00$	$+24.58$	-3.10	-13.77
	12.73	High	$+154.60$	-36.14	-32.80	$+9.08$	$+70.83$
$Y_{NH_3\,\text{③}}$ / mol %	3.39	Low	-18.64	$+5.99$	$+6.22$	-1.50	-7.07
	4.53	High	$+8.85$	-10.00	-10.26	$+2.51$	$+13.91$
H_2:N_2 ③ / mol of H_2 : mol of N_2	1.18 : 2.82	Low	-86.00	-67.15	-73.39	$+17.19$	$+256.80$
	3.05 : 0.95	High	$+7.01$	-5.99	-6.30	$+1.62$	$+8.64$
$\dot{m}_{\text{③}}$ / kg h^{-1}	527.78	Low	-20.33	-16.26	-16.16	-21.36	-6.08
	707.61	High	$+6.80$	-3.00	-3.22	$+9.26$	$+12.65$
$T_{\text{③}}$ / K	519.41	Low	-0.68	-7.76	-7.94	$+1.95$	$+10.53$
	536.84	High	$+2.64$	-1.82	-1.86	$+0.45$	$+2.32$

rate, thus temperature also increases and the temperature in bed 1 reaches the upper limit of catalyst *i.e.* 803 K. Therefore the reactor cannot be operated beyond 213.91 bar, although the reactor system is capable of autothermic operation greater than 213.91 bar. Increase in pressure provides more flexibility in operation and NH_3 production than decrease in pressure, but at the expense of more H_2 consumption, see table 2.3.

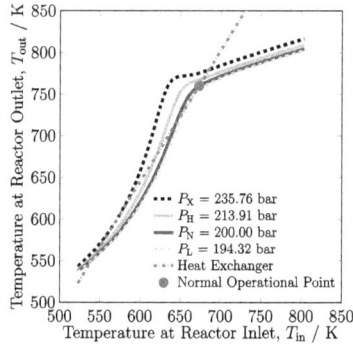

Figure 2.4.: Steady-state characteristics of the reactor system for highest (X), high (H), normal (N) and low (L) operational pressures of the reactor system.

The pressure dependence of the outlet temperature is given in figure 2.5. The stable steady state points are covered by the solid line and unstable steady state points by dotted line. The stable operational envelope for pressure is 194.32 to 213.91 bar. Decreasing the inlet temperature at bed 1 or pressure within the reactor system below *ca.* 663 K or 194.32 bar leads to the

reactor system shutdown, and increasing inlet temperature at bed 1 or reactor pressure above

679 K or 213.91 bar results in an exit gas temperature greater than 803 K for catalyst bed 1. In

the given pressure range, multiple states are possible and due to this multiplicity the branch

switching is also possible. The upper branch is desired for stable steady state operation.

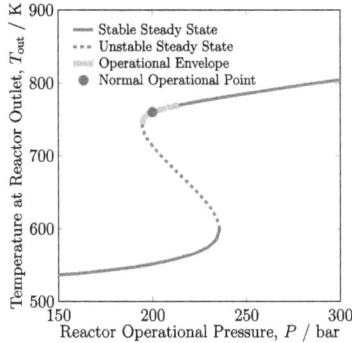

Figure 2.5.: Steady state characteristics of the reactor system for outlet temperature *versus* operational pressure of the reactor system.

The dependence on the stable operating range of the autothermic reactor system on the inert

gas concentration in feed is shown in figure 2.6. The exit gas temperature of the reactor system

decreases by 30 K, *i.e.* from 760 to 730 K with addition of inert gas in the feed. Temperature

of the exit gas increases to *ca.* 770 K with removal of inert gas in the feed, see table B.2,

appendix B. The underlying reason is that reactant concentration decreases or increases with

addition or removal of inert gas in the feed, respectively. Furthermore, as can be evident from

table 2.3, with increase and decrease in inert gas concentration in feed, a H_2 intake decreases

and increases in feed by 36.14 % and 15.00 %, respectively. A maximum operating envelope

of 0 to 12.73 mol % inert species is identified. Here, 0 mol % of inert gas means zero purging

of gas from recycle stream and fresh stream ① consist of H_2 and N_2 only. Inert gas higher

than 12.73 mol % is not suitable for autothermic operation of the reactor system, as the heat

of removal will be greater than the heat produced by ammonia synthesis reaction.

In figure 2.7, outlet temperature *versus* ammonia concentration in the feed for the reactor sys-

tem (stream ③) is shown. The reverse S-shaped curve presents up to three steady state points

in the range of 2.84 to 4.53 mol % ammonia concentration in feed. The desired operational

envelope for ammonia concentration in the feed is quite narrow with 3.39 to 4.53 mol %. The

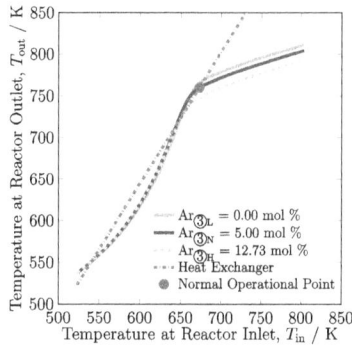

Figure 2.6.: Steady-state characteristics of the reactor system for low (L), normal (N) and high (H) argon (inert gas) concentrations in feed ③ of the reactor system.

switching of the branch above $4.53\,NH_3$ mol % in feed results in reactor operation instability, and operating below $3.39\,NH_3$ mol % results in temperature higher than the catalyst sustainability limit in catalyst bed 1, see table B.2, appendix B. A decrease in ammonia concentration in the reactor feed results in higher outlet temperature and higher reactants conversion by 8 K and 1 %, respectively from normal operation. The load on the recycle stream is reduced slightly by 1.5 %, at the expense of 6 % more hydrogen consumption, also see table 2.3. Whereas, with an increase in ammonia concentration in the reactor system intake, reactants composition decreases, and results in lower conversion and temperature rise in all catalyst beds.

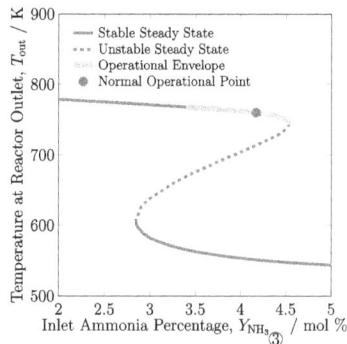

Figure 2.7.: Steady state characteristics of the reactor system for outlet temperature *versus* ammonia concentration in feed ③ of the reactor system.

The operational envelope for the H_2-to-N_2 ratio is quite wide for autothermal operation of the reactor system, which is evident from figure 2.8. The reactor can be operated for H_2-

to-N_2 ratios between 1.18 : 2.82 and 3.05 : 0.95. However, operating the reactor under a non-stoichiometric ratio noticeably reduces H_2 intake and increases the recycle load, see table 2.3. For the reactor system operation under a non-stoichiometric ratio of reactants, the feed stream ① composition also varies from the nominal value, and new compositions are calculated by using equations A.32 to A.34, see appendix A. The reactor at H_2-to-N_2 ratio of 1.18 to 2.82 (H_2 is limiting reactant) and 3.05 to 0.95 (N_2 is limiting reactant) results in *ca.* 37.5 and 22 % of H_2 conversion, and *ca.* 5 and 23.5 % of N_2 conversion, respectively, compare to *ca.* 24.5 % of reactants for normal operation. Also, it should be noted that the reactor temperature decreases by up to 90 K with decrease in H_2-to-N_2 ratio and enhances limited reactant conversion, see table B.2, appendix B. The operation of the reactor system at a ratio other than 3 mol of H_2 to 1 mol of N_2 reduces NH_3 production. But the low H_2-to-N_2 ratio, which corresponds to a lower hydrogen intake, is still beneficial during renewable power, *i.e.* hydrogen production outage for small period of time, as it will not let the ammonia synthesis reactor blow out. As such, the H_2-to-N_2 ratio may be a major manipulable for renewable energy availability based control of such plants.

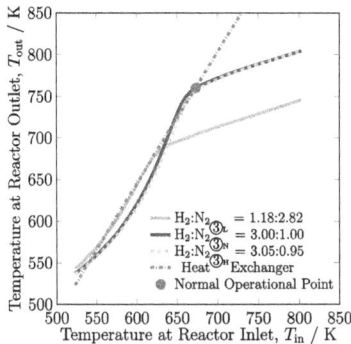

Figure 2.8.: Steady-state characteristics of the reactor system for low (L), normal (N) and high (H) H_2-to-N_2 ratios in feed ③ of the reactor system.

To adjust for fluctuation of renewables, total feed flow inlet may be adjusted. The maximum and minimum total feed flow rates are 707.61 to 527.78 kg/h respectively, with corresponding ammonia productions of 116.13 and 100.60 kg/h. The change in total feed flow rate is realised by a proportional change in quenches. A decrease in total flow rate results in a decline in the hydrogen intake by *ca.* 16 % and in recycle load by *ca.* 16 %. On the other hand, significant increase in total flow rate was not possible, and therefore not much change in hydrogen intake and recycle load occurred, see table 2.3. The exit temperature (see figure 2.9) and overall conversion of the reactor remains higher for flow rates below the normal total feed flow rate and vice versa, see also table B.2, appendix B. This is due to the fact that the reaction reaches equilibrium conditions well before exiting from bed 3 at lower flow rates. Whereas, with increase in flow rate, the space velocity also increases and it results in lower rate of reaction. Like for other process variables, the operating envelope for total feed flow rate also lies inside the multiplicity region, and it is again limited by stability of the reactor system and maximum catalyst temperature in bed 1.

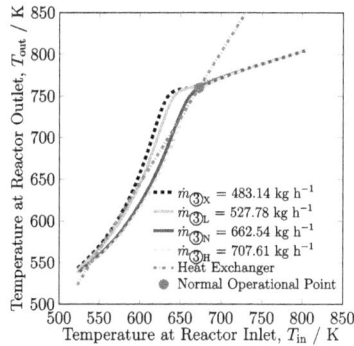

Figure 2.9.: Steady-state characteristics of the reactor system for lowest (X), low (L), normal (N) and high (H) total feed ③ flow rates of the reactor system.

Changing the feed temperature entering the reactor system changes not only the heat production curve but also the heat removal line. The feed temperature influences the location of both the curve and the line in the opposite direction: with the increase in feed temperature, the heat production curve moves upwards, while the heat removal line moves downwards, as can be seen in figure 2.10. This distinguishes feed temperature from the other investigated process variables; with changes in feed temperature, the y-intercept of heat removal curve also changes, see equation 2.1. The operating envelope for the feed temperature is between 519.41 and 536.84 K, where from 519.41 to 536.08 K lies inside the multiplicity region, and above 536.08 K the heat production curve intersects the heat removal line at only one point. The minimum and maximum limit of feed temperature is set due to stability of the reactor system and maximum temperature reached in catalyst bed 1, respectively. Operation of the reactor system at conditions other than normal feed temperature i.e. 523 K, reduces H_2 intake up to ca. 8 % and NH_3 production up to ca. 8 % at the expense of a slight increase of recycle load up to *ca.* 2 %, see table 2.3. Overall, change in the feed temperature results in a decline in conversion from normal operation, see table B.2, appendix B. Whereas, it can be seen that for higher feed temperature, conversion in bed 1 and 2 is higher from normal operation, but conversion in bed 3 is lower, which is attributed to higher temperature operation *i.e.* equilibrium is approached before exit of bed 3.

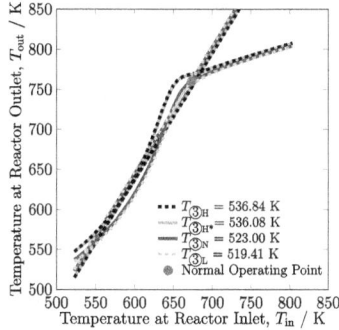

Figure 2.10.: Steady-state characteristics of the reactor system for high (one (H) and two intersections (H*)), normal (N) and low (L) feed ③ temperatures of the reactor system.

After comparing results for process variables from figures 2.4 to 2.10, table 2.3 and table B.2 (appendix B) it is concluded that by reducing H_2-to-N_2 ratio, increasing inert gas concentration and decreasing feed flow rate have the most potential to reduce the H_2 consumption by up to *ca.* 67 %, 36 % and 16 %, respectively. This decrease in H_2 intake comes along with variations in recycle load; with H_2-to-N_2 ratio reduction and inert gas concentration increase, the recycle load increases by 17 % and 9 %, respectively and along with decrease in feed flow rate the recycle load also decreases. Among the six process variables, inert gas concentration in the feed provides the maximum operational flexibility, almost increasing by 255 % from the normal value, and without inert gas in the synthesis loop, H_2 consumption increases by 15 %. The other three process variables barely impact H_2 consumption (below 10 %) and recycle load (below 3 %), see table 2.3. The higher temperature operational limit of 803 K is approached in catalyst bed 1 at a lower boundary of NH_3 and feed flow rate, and at an upper boundary of pressure and feed temperature.

2.4. Conclusions

This chapter presented a systematic analysis of the operating and production flexibility of a Haber-Bosch ammonia reactor. From the results, it can be concluded that the autothermic reactor is viable for power-to-ammonia process, as it can be operated for a wide range of process variables while maintaining operational, hydrogen feed intake and ammonia production flexibilities. Operating outside these boundaries leads to the shutdown of reactor system autothermic operation or damage to the catalyst due to overheating. Among the six process variables, H_2-to-N_2 ratio and inert gas concentration in the reactor system feed provide the most flexibilities with up to *ca.* 67 % decrease in H_2 intake. This state may be advantageous to prevent the production plant from shutting down during phases of low availability of the H_2 produced from the renewables. Further, it can be noticed that changes in H_2-to-N_2 ratio and feed temperature from the nominal operational values result in a decline in hydrogen intake and ammonia production, causing the load on recycle stream to increase, whereas higher temperature operational limit is always reached in the catalyst bed 1. This study showed that despite present Haber-Bosch reactors being operated only at their optimum, the reactor system is feasible to operate over a wide load range, and is thus attractive for power-to-ammonia applications.

3. Performance comparison of autothermal reactor systems

The reactor system design has a strong influence on the rate of ammonia production; therefore in this chapter, a flexibility analysis for various configurations of the autothermal synthesis reactor system is made. In contrast to chapter 2, the flexibility analysis is performed by optimisation of ammonia production by varying one process variable at a time. Also, instead of six process variables, only three variables (H_2-to-N_2 ratio, inert gas concentration and process feed flow rate) are considered, as they showed to provide higher operational and production flexibilities. Through minimisation and maximisation of NH_3 production, the load range of synthesis reactor systems is enquired and compared. This chapter is organised as follows: first, literature review of multi-bed ammonia synthesis reactors along with their optimisation strategies is carried out. Afterwards, possible reactor system configurations along with adapted optimisation strategies for performing flexibility analysis are presented. At the end, results are discussed and concluding remarks are made.

3.1. Ammonia synthesis reactor systems

From chapter 1 it is already known that three-bed reactor systems are economical and efficient reactor systems. Furthermore, various inter-stage cooling configurations can be made e.g. direct and indirect cooling. In addition, three-bed synthesis reactor systems also work with hybrid inter-stage cooling configuration, i.e. inter-cooled and directly cooled, Farivar & Ebrahim[46] also proposed a new configuration consisting of direct and indirect cooling reactor systems combination, which results in increased N_2 conversion. In their proposed configuration, they introduced direct cooling between bed 1 and 2, and indirect cooling with the help of the feed between beds 2 and 3.

A large amount of optimisation work on multi-bed reactor systems can also be found.[45,47,55–57] A four-bed direct cooling reactor system for optimal operational temperature by a one-dimensional homogeneous model was presented by Gaines[55], whereas an indirect cooling reactor system for maximum N_2 conversion by a one-dimensional homogeneous model was analysed by Aka & Rapheel[47]. Gaines optimisation results show that with good temperature control, NH_3 production can be increased by 1 % for constant feed rate. Also, impact of quench fractions on reactor efficiency, either alone or with one process variable at a time, were studied.[55] Aka & Rapheel considered each bed inlet temperature as the decision variable for maximising N_2 conversion from 0.19 to 0.26.[47] Elnashaie et al.[56], Elnashaie & Alhabdan[57] and Azarhoosh et al.[45] optimised three-bed indirect cooling reactor systems for maximum NH_3 production by using a one-dimensional heterogeneous model. Elnashaie et al. determined the optimal temperature profile and increased NH_3 production by 6 to 7 % for reactor operation at optimal temperature profile. Furthermore, their model satisfied various adiabatic and non-adiabatic industrial reactor systems.[56] Elnashaie & Alhabdan developed a computer software to determine the optimal behaviour of an ammonia reactor, whence three internal collocation points were required for achieving satisfactory results.[57] Azarhoosh et al. used a genetic algorithm for optimisation of inter-cooled and directly cooled horizontal reactors for maximum ammonia production. They considered an inter-cooled reactor with only one heat exchanger between beds 1 and 2, exchanging heat with the reactor feed. Furthermore, the effect of parameters such as inlet temperature, total feed flow rate and operation pressure on NH_3 production were studied. Satisfactory matching of the results between simulation and industrial data was achieved.[45] In a most recent optimisation work for maximising N_2 conversion by differential evolution algorithm, Khadmi & Sabbaghi also compared adiabatic direct and indirect cooling reactor systems.[34] They considered a one-dimensional pseudo-homogeneous model along with an empirical relation for diffusion resistance. The indirect cooling reactor system results in higher N_2 conversion with only the inlet temperature of each bed as the decision variable.

In addition to optimum design, as well as efficient and autothermal operation of ammonia synthesis reactor system, power-to-ammonia also requires operational and production flexibilities due to dependency on intermittent renewable energy. In chapter 2 the operating envelope of a three-bed autothermal reactor system was determined via a one-dimensional pseudo-homogeneous model and steady-state stability analysis for six variables. Among six process variables, inert gas fraction, H_2-to-N_2 ratio and total feed flow rate provided significantly higher operational flexibilities from nominal value. Operation of the reactor system with variables other than nominal H_2-to-N_2 ratio or total feed flow rate resulted in lower ammonia production.

It can be seen that numerous studies on optimisation of reactor system configurations, max-imising N_2 conversion and NH_3 production have been carried out. However, no significant work for off-design performance analysis, *i.e.* minimisation and maximisation of NH_3 produc-tion has been done. Raw materials are abundantly available for conventional ammonia pro-duction; therefore minimisation of NH_3 production was not of interest. Whereas for power-to-ammonia, besides design performance, off-design performance is of equal importance, as reactants supply is strongly dependent on supply of renewable energy. For example, H_2 pro-duction *via* electrolysers consume more than 90 % of the energy.[24] Therefore during energy shortage periods, H_2 production needs to be shut or slowed down. Thus, the focus of this chapter is to investigate the ammonia production flexibilities, along with H_2 intake, change in recycle load and recycle to feed ratio by manipulating the suitable process variables (inert gas fraction, H_2-to-N_2 ratio and total feed flow rate) for three-bed additional autothermal reactor system configurations. Therefore, first the reactor system configurations and the synthesis loop are defined. Afterwards, problem formulation for optimisation case scenarios: design performance and off-design performance (minimum and maximum NH_3 production) for flex-ibility analysis is made. Finally, results are compared and discussed.

3.2. Methodology

In addition to efficient and autothermal operation, power-to-ammonia requires operational and production flexibilities. Therefore, special emphasis should be given to the selection of a reactor system that is capable of maintaining autothermal operation over a wide range of NH_3 production rates. In this section, first the three-bed autothermal reactor system con-figuration is discussed, as beside efficient operation, it also allows various combinations of interstage cooling methods for optimal heat management as shown in figure 3.1. The configu-rations, shown in figures 3.1a, 3.1d and 3.1e, represent standard quench type cooling (2Q) and inter-stage cooler based cooling (2H-2 and 2H-3) synthesis reactor systems, respectively. The configurations shown in figures 3.1b and 3.1c are tailor made synthesis reactor systems, with a combination of quench-type and inter-stage cooler based cooling before (HQ) and afterwards (QH), respectively. Determination of the most suitable autothermal reactor systems among the design variants for power-to-ammonia would be of great interest.

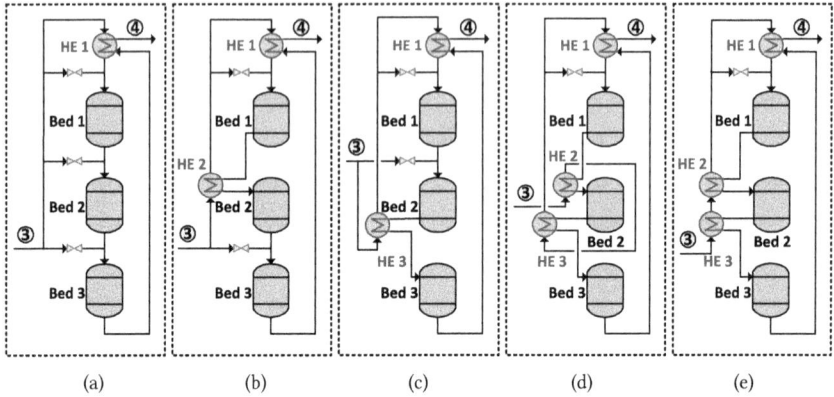

(a)   (b)   (c)   (d)   (e)

Figure 3.1.: Three-bed autothermal ammonia synthesis reactor systems: (a) direct cooling by quenching (2Q); (b) combination of indirect & direct cooling (HQ); (c) combination of direct & indirect cooling (QH); (d) indirect cooling by heat exchange between process streams with feed entering first HE 2 (2H-2); (e) indirect cooling by heat exchange between process streams with feed entering first HE 3 (2H-3).

To compare the performance of the synthesis reactor systems 2Q, HQ, QH, 2H-2 and 2H-3, and their interaction with the synthesis loop, the Haber-Bosch ammonia synthesis loop is divided into two boundaries, see figure 3.2. The system boundary I is applied around a synthesis reactor system and the system boundary II is applied around the overall synthesis loop. However, to keep the model generic and focused on the complex reactor systems (boundary I) and their minimum and maximum NH_3 production limitations, the limitations which may occur by the unit operations in the system boundary II are neglected, similar to chapter 2. For making comparison among the design variants, two performance scenarios are considered: design and off-design. Design performance analysis is made at stable process conditions for a normal NH_3 production load, and off-design performance analysis is made by varying one process variable at a time for minimum and maximum NH_3 production load. For design and off-design performance comparison among the design variants, it is necessary that it is made on equal grounds, *i.e.* at similar operational conditions, process feed composition and flow rate. For performance comparison among the design variants, most of the model from chapter 2 is adopted. However, for the heat exchangers in the reactor systems model, the assumptions remain similar to chapter 2, but a more generalised model is presented. In this section, first the mathematical models required by the system boundaries I and II are presented along with optimum design conditions for normal NH_3 production load. Afterwards, off-design operation scenarios are discussed with variation of one process variable at a time.

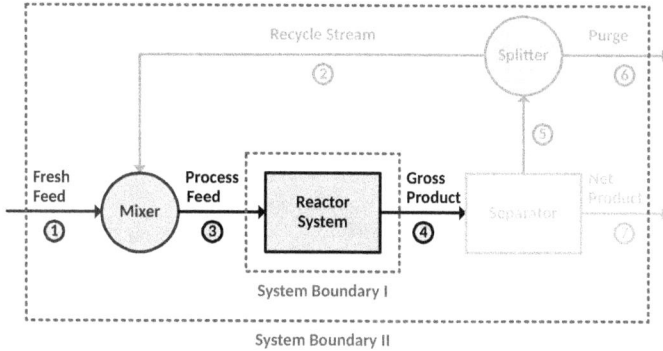

Figure 3.2.: Haber-Bosch ammonia synthesis loop along with system boundaries I and II.

3.2.1. Mathematical model

For the comparison of the design and off-design performance among the design variants (figure 3.1), the models are kept compact, with as little detail as necessary. The assumptions and mathematical models applied within the system boundaries I and II are given as follows.

System boundary I

To analyse the synthesis reactor systems, a given synthesis reactor system is divided into three individual sub-systems: catalyst bed, mixer and heat exchanger. By combining the model of each individual subsystem, the behaviour of a given overall synthesis reactor system is quantified.

For the catalyst beds and the inter-stage mixers, the model is taken from chapter 2. The catalyst bed is described by the steady-state one-dimensional pseudo-homogeneous species and energy balance equations 2.2 and 2.3. The model for the mixer is described by the steady-state species and energy balance equations 2.5 and 2.6. For the heat exchanger, the model assumptions are similar to chapter 2, but equation 2.1 is customised for reactor system 2Q only. Here, with regard to several design variants, a more general form is needed. Therefore, for the previously stated assumptions, the heat exchanger energy balance can be described by an ε-NTU model [48], equations 3.1a and 3.1b:

$$\varepsilon_{HE,RS} = \frac{Q_{HE,RS}}{Q_{MAX,HE,RS}} = \begin{cases} \text{For } (\dot{m}C_p)_{\text{hot}} < (\dot{m}C_p)_{\text{cold}} : \\ \dfrac{(\dot{m}C_p)_{\text{hot},HE,RS} \left(T_{\text{hot, in}} - T_{\text{hot, out}}\right)_{HE,RS}}{(\dot{m}C_p)_{\text{MIN},HE,RS} \left(T_{\text{hot, in}} - T_{\text{cold, in}}\right)_{HE,RS}} \qquad (3.1a) \\[1em] \text{For } (\dot{m}C_p)_{\text{cold}} < (\dot{m}C_p)_{\text{hot}} : \\ \dfrac{(\dot{m}C_p)_{\text{cold},HE,RS} \left(T_{\text{cold, out}} - T_{\text{cold, in}}\right)_{HE,RS}}{(\dot{m}C_p)_{\text{MIN},HE,RS} \left(T_{\text{hot, in}} - T_{\text{cold, in}}\right)_{HE,RS}} \qquad (3.1b) \end{cases}$$

where $\varepsilon_{HE,RS}$ is the effectiveness of heat exchanger $HE \in \{1,2,3\}$ of respective reactor system $RS \in \{$2Q, HQ, QH, 2H-2, 2H-3$\}$. $Q_{HE,RS}$ is heat transfer rate, $T_{\text{hot},HE,RS}$ is the temperature at tube side, $T_{\text{cold},HE,RS}$ is the temperature at shell side, $\dot{m}$ is the mass flow rate and C_p is the specific heat of process gas at respective side (hot and/or cold) of heat exchanger. In equation 3.1a, $(\dot{m}C_p)_{\text{MIN}} = (\dot{m}C_p)_{\text{hot}}$ and in equation 3.1b, $(\dot{m}C_p)_{\text{MIN}} = (\dot{m}C_p)_{\text{cold}}$. The additional equations for calculating NTU and the area of a shell and tube heat exchanger are mentioned in table B.3, appendix B.

System boundary II

For acquiring design and off-design performance among the design variants, it is necessary to extend the model from the reactor systems to the overall synthesis loop, as the recycle stream influences process feed composition, see figure 3.2. The implemented material balance equations are provided in section A.2 (appendix A). Herein, by solving equations A.32 to A.34 simultaneously, the purge ratio p_{RS}, the mass fraction of the N_2 in the fresh feed $x_{N_2,①,RS}$ and the mass flow rate of the fresh feed $\dot{m}_{①,RS}$ are evaluated for all the design variants. In addition, equations A.23, A.27, A.21 and A.22 are used to calculate components $c \in \{N_2, H_2, Ar, NH_3\}$ mass flow rate $\dot{m}_c$ in streams $⑤ \in \{5,2,6\}$, and by a sum of the components mass flow rates in the respective stream, the overall flow rate of the stream is evaluated. Finally, by using equation A.35, net product rate $\dot{m}_⑦$ is calculated.

For off-design performance cases, flexibility analysis is also performed for the reactor systems with loop. The flexibility for NH_3 production $\dot{m}_{⑦,RS}$, H_2 intake $\dot{m}_{H_2,①,RS}$, operational for process variables, recycle $\dot{m}_{②,RS}$ and recycle to feed ratio $(\dot{m}_{②,RS}/\dot{m}_{①,RS})$ is defined as the relative change from the normal values and calculated by using equation 2.7.

3.2.2. Problem formulation

For the design variants' performance analysis, the optimisation problem is developed for design and off-design scenarios with consideration of 2Q as a reference system.

Design performance

To compare the design variants 2Q, HQ, QH, 2H-2 and 2H-3 at their nominal operation point, parameters and process variables are kept constant for each system boundary. The compositions of fresh and process feed, along with the operation conditions are taken from table 2.1 for a respective system boundary.

System boundary I

The catalyst bed volumes and the process feed flow rate of reactor system 2Q are taken from table 2.2 for other design variants and summarised in table 3.1.

Table 3.1.: Process feed flow rate and catalyst bed volumes

Flow rate of process feed	Volume of catalyst beds			
$\dot{m}_{\circled{3}}$ / kg h^{-1}	V_{b1} / m^3	V_{b2} / m^3	V_{b3} / m^3	$V_{b\,Total}$ / m^3
662.54	0.0075	0.0221	0.0464	0.0725

To determine the maximum gross NH_3 production of the respective reactor system, equality (equations 2.2 to 2.6) and inequality constraints (equations 3.2a and 3.2b) are implemented to assure that all catalyst beds operate at maximum temperatures.

$$\underset{z_{RS}}{\text{maximise}} \ \dot{\mu}_{NH_3,RS} = \dot{m}_{NH_3,\circled{4},RS} - \dot{m}_{NH_3,\circled{3},RS}$$

subject to

Equations 2.2 to 2.6 (reactor system).

$$673\,K \leq T_{b_{in},RS} < T_{b_{out},RS} = \begin{cases} 773\,K; \ \text{for } b \in \{1,2\} & (3.2a) \\ 0.90\,T_{EQ,RS}; \ \text{for } b = 3 & (3.2b) \end{cases}$$

where $\dot{\mu}_{NH_3}$ is the gross NH_3 production for the reactor systems $RS \in \{HQ, QH, 2H\text{-}2, 2H\text{-}3\}$. For reactor systems HQ, QH and 2H, inlet temperatures of the catalyst beds are manipulated by means of direct mixing and/or inter-stage heat exchangers, see figure 3.1. For the reactor systems having an interstage mixer, heat exchanger or combination of both, the process variables are quench flow rate, catalyst bed inlet temperature or both; thus z_{RS} of respective reactor systems are $z_{HQ} \in \{T_{b1_{in}}, T_{b2_{in}}, \dot{m}_{q3_{in}}\}$, $z_{QH} \in \{T_{b1_{in}}, \dot{m}_{q2_{in}}, T_{b3_{in}}\}$ and $z_{2H} \in \{T_{b1_{in}}, T_{b2_{in}}, T_{b3_{in}}\}$. For the catalyst beds of each reactor system, $T_{b_{out},RS}$ is the outlet temperature limited by the constraint. For bed $b \in \{1, 2\}$, $T_{b_{out},RS}$ is maintained to $773\,\mathrm{K}$ and for bed 3, exit temperature is maintained to 90 % of the equilibrium temperature $T_{EQ,RS}$ to limit the catalyst bed size.[37] The effectiveness $\varepsilon_{HE,RS}$ of heat exchangers $HE \in \{1, 2, 3\}$ of the respective reactor system are calculated by equations 3.1a and/or 3.1b at optimum design conditions, and later used in the off-design performance analysis for identifying the unknown stream temperatures.

System boundary II

The initial composition of the fresh feed stream ① is taken from table 2.1 and kept constant for all the design variants (see figure 3.1), so performance of the ammonia synthesis loop shown in figure 3.2 can be made by using relevant equations, mentioned in section 3.2.1.

Off-design performance

It is pertinent to mention that the design variants shown in figure 3.1 are designed in consideration of autothermal operation, therefore it is also necessary that for off-design performance analysis, the energy balance between heat of production and heat of consumption should be retained. For this, the van Heerden[39] steady state stability approach has been adapted, as previously discussed in section 2.2.2.

System boundary I

For the identification of the reactor systems' minimum and maximum gross ammonia production, in addition to equations 2.2 to 2.6, equality and inequality constraints equations 3.1a, 3.1b, 3.3 and 3.4 are also implemented. This yields the following optimisation problem.

$$\underset{z_{③,RS}}{\text{minimise} \mid \text{maximise}} \; \dot{\mu}_{NH_3,RS} = \dot{m}_{NH_3,④,RS} - \dot{m}_{NH_3,③,RS}$$

subject to

Equations 2.2 to 2.6 (reactor system).

Equation(s)3.1a and/or 3.1b (reactor system).

$$623\,K \leq T_{b_{in},RS} < T_{b_{out},RS} \leq 803\,K \mid T_{EQ,RS} \tag{3.3}$$

$$T_{b1_{in},RS} + \Delta T_{min} \leq T_{b3_{out},RS} \tag{3.4}$$

where $\dot{\mu}_{NH_3}$ is the gross NH_3 production for reactor systems $RS \in \{2Q, HQ, QH, 2H\}$. All reactor systems have the following process variables: inert gas concentration in process feed, process feed mass flow rate and reactants ratio in process feed $z_{③,RS} \in \{y_{Ar} \mid \dot{m} \mid y_{H_2} : y_{N_2}\}_{③,RS}$. For comparing their single impact on ammonia production flexibility, only one process variable is manipulated and the others are fixed. Furthermore, $T_{b1_{in}}$ is set free for all the reactor systems (and in addition $T_{cold_{out}, HE3}$ for reactor system 2H-3 only), and the value is selected by the optimiser itself with respect to constraints. In comparison to the design performance analysis, for off-design performance analysis the operational temperature range of catalyst beds can be increased by up to 80 K, see equation 3.3. For example, for iron-based catalyst upper temperature limit is 803 K and the lowest temperature is 623 K below which the reaction rate is too low.[6] In all the heat exchangers of the reactor systems, a minimum temperature difference is necessary to guarantee sufficient heat transfer between two streams. Usually, the practical minimum temperature difference is between 10 and 20 K[58]; here for $HE_{RS} \in \{1, 2, 3\}_{RS}$ a $\Delta T_{min} = 20\,K$ limitation is implemented.

System boundary II

Similarly to design performance, the material balance of the process streams of ammonia synthesis loop for off-design performance is performed. However, for the maximisation of NH_3 without Ar gas presence in the synthesis loop, the fresh feed gas is also considered free of Ar gas. As for the higher operational cost, pure N_2 is achievable by pressure swing adsorption or cryogenic distillation.[23]

3.2.3. Problem implementation

The simulations and optimisation problems are performed in MATLAB, where the system of equation is solved by the optimiser fmincon for system boundary I, and for the implementation of differential equations related to catalyst beds, the ODE solver (ode45) is called. For initialising the optimisation problem, initial guesses of design parameters and/or process variables are provided. Afterwards, the optimisation routine continues iteratively until the constraint conditions are met. As fmincon is a local optimiser, to check the credibility of the optimum solution the optimisation problem was repeatedly performed for multiple initial guesses of design parameters and process variables.

The flow charts related to the design and off-design performance optimisation problems are illustrated in figure 3.3. It can be seen that the optimiser solves the problem for the system boundary I and afterwards transfers the optimised values to system boundary II. To evaluate the overall material balance accordingly.

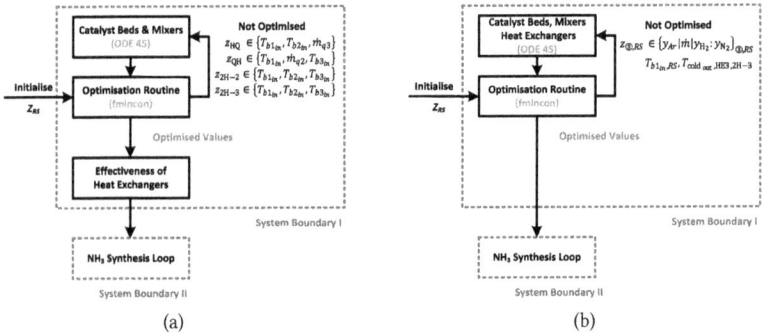

Figure 3.3.: Flow chart of the optimisation problems (a) design performance and (b) off-design performance.

3.3. Results and discussions

The results obtained for all design variants are presented and discussed in this section. First, reactor systems and synthesis loop design performance analysis is performed for maximum NH_3 production, afterwards off-design performance analysis is performed for minimum and maximum NH_3 production by varying one process variable at a time. In off-design performance analysis, in addition to minimising and maximising NH_3 production, stability analysis also needs to be performed for the variation of the three process variables; inert gas concentration, process feed flow rate and reactants ratio in fresh feed stream. Furthermore, flexibilities of NH_3 production, H_2 intake, process variable operation, recycle load and recycle to feed stream ratio for off-design performance of all design variants are analysed.

3.3.1. Design performance

In this section, results obtained for the maximum NH_3 production from the various reactor systems (system boundary I) and the synthesis loop (system boundary II) for the same initial conditions are presented and discussed. First, the optima for the three variables for the reactor systems $RS \in \{2Q, HQ, QH, 2H\text{-}2, 2H\text{-}3\}$ are presented and discussed, followed by temperature-reactants conversion $\overline{TX}$ trajectories. Afterwards, the effectiveness of heat exchangers in reactor systems and the flow distribution within the synthesis loop are analysed at the operational point of maximum NH_3 production.

System boundary I

The maximum gross NH_3 production resulting from the optimisation of the process variables inlet temperature of catalyst beds and/or quench stream flow rates are given in table 3.2. It can be observed that reactor systems 2H-2 and 2H-3 result in the highest gross NH_3 production. The primary reason for NH_3 production difference among the reactor systems is their different space time: the higher the space time, the higher the reactants conversion. For example, reactor systems 2H-2 and 2H-3 are the systems where the whole process feed passes through all three catalyst beds, therefore resulting in the highest reactants conversion.

Table 3.2.: Optimal process variables (highlighted in grey colour) for the reactor systems and resulting maximum ammonia production rate (enclosed in rectangle)

RS	$T_{b1_{in, NOR}}$ / K	$\dot{m}_{q2_{NOR}}$ / kg h^{-1}	$T_{b2_{in, NOR}}$ / K	$\dot{m}_{q3_{NOR}}$ / kg h^{-1}	$T_{b3_{in, NOR}}$ / K	$\dot{\mu}_{NH_3, NOR}$ / kg h^{-1}
2Q[1]	673.00	163.83	673.00	177.01	673.00	121.11
HQ	691.14	-	681.01	166.74	675.07	123.70
QH	687.63	207.89	683.06	-	678.87	129.08
2H-2	702.39	-	688.16	-	680.30	131.38
2H-3	702.39	-	688.16	-	680.30	131.38

Temperature progression and reactants conversion at a stoichiometric ratio along with the equilibrium line for optimum operation of the reactor systems are presented in figure 3.4. During design performance, reactor systems 2H-2 and 2H-3 resulted in the same optimum values, and therefore are represented as a one system i.e. 2H. As the stoichiometric ratio of H_2-to-N_2 is fixed, their conversion profiles overlap each other and therefore are represented with

---

[1] Optimal values of process variables and ammonia production rate for the reactor system 2Q are taken from chapter 2.

one line only. Due to the exothermic reaction, temperature increases along with the reactants conversion in the catalyst beds. All the reactor systems are operated at the maximum possible temperature or reactants conversion, which is evident from figure 3.4, as for beds 1 and 2 maximum temperature and for bed three maximum conversion is reached. Overall, reactor systems 2H-2 and 2H-3 (2H) result in the highest reactants conversions of 26.77 %, whereas in the reactor systems QH, HQ and 2Q achieve 0.47 %, 1.56 % and 2.09 % less conversion, respectively.

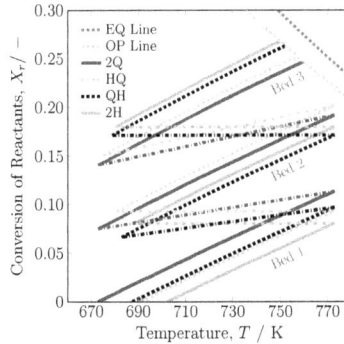

Figure 3.4.: Temperature-reactants conversion $\overline{TX}$ trajectories for reactor systems 2Q, HQ, QH and 2H (2H-2 and 2H-3) along with equilibrium (EQ) and operational (OP) lines.

It can be observed from figure 3.4 that for heat exchanger and quench-based cooling, the temperature-reactants conversion $\overline{TX}$ trajectory shifts away from the equilibrium. Furthermore, the impact of the interstage heat exchangers and mixers (dash dotted lines) can also be differentiated, whereas for the heat exchanger-based cooling, only temperature decreases and for the quench-based cooling, reactants conversion also decreases.

The effectiveness of the heat exchangers for optimum NH_3 production by reactor systems 2Q, HQ, QH, 2H-2 and 2H-3 is presented in table 3.3. The effectiveness of heat exchangers is calculated for the optimum design performance of all reactor systems by either equation 3.1a or 3.1b. Therein, equation 3.1a is applied to all three heat exchangers of reactor system 2H and heat exchanger 2 of reactor system HQ, and equation 3.1b is applied to both heat exchangers of reactor system QH and heat exchanger 1 of reactor systems 2Q and HQ.

Later, for off-design performance analysis, the above mentioned effectivenesses of the heat exchangers are used for calculating the temperature of a required stream.

Table 3.3.: Resulting effectiveness of heat exchangers for optimised reactor systems

RS	$\varepsilon_{HE\,1}$ / -	$\varepsilon_{HE\,2}$ / -	$\varepsilon_{HE\,3}$ / -
$2Q^2$	0.6329	-	-
HQ	0.5675	0.3680	-
QH	0.4054	-	0.4821
2H-2	0.3398	0.3393	0.5391
2H-3	0.3398	0.5007	0.3708

System boundary II

The identified optimal temperatures and feed distributions for the various reactor systems are now integrated in the synthesis loop, and the flow distribution for the overall synthesis loop is calculated. The overall ammonia synthesis loop material balance is performed by the equations stated in section A.2 of appendix A and the summary is presented in table 3.4. It can be observed that with an increase in the fresh feed intake ①, ammonia production ⑦ also increases and recycle stream ② load decreases. The reactor systems 2H and QH are able to produce $ca.$ 10 and 8 kg/h more NH_3 than the reactor system 2Q. However, this higher NH_3 production requires $ca.$ 11 and 9 kg/h more fresh feed intake than reactor system 2Q. Furthermore, an increase in fresh feed ① also increases the amount of inert in the synthesis loop; as a consequence purge ⑥ also increases by 8 %. With increase in NH_3 production, a more efficient separator is also required, to maintain a constant process feed ③ composition.

Table 3.4.: Flow distribution in synthesis loop for maximum NH_3 production by varying temperatures and feed distributions in the catalyst beds of the reactor systems

RS	Fresh Feed $\dot{m}_①$	Purge Percent. $p = \dfrac{\dot{m}_⑥}{\dot{m}_⑤} \times 100$	Purge $\dot{m}_⑥$	Recycle $\dot{m}_②$	Product $\dot{m}_⑦$	$\eta_{NH_3} = \dfrac{\dot{m}_⑦}{\dot{m}_{NH_3,④}} \times 100$
	/ kg h^{-1}	/ %	/ kg h^{-1}	/ kg h^{-1}	/ kg h^{-1}	/ %
$2Q^3$	133.05	2.41	13.05	529.48	120.00	72.21
HQ	135.82	2.46	13.26	526.70	122.56	72.63
QH	141.58	2.56	13.69	520.95	127.90	73.44
2H-2	144.04	2.60	13.87	518.49	130.17	73.78
2H-3	144.04	2.60	13.87	518.49	130.17	73.78

All in all, all five reactor systems show relatively similar performance with less than 10 % deviations from each other. They are thus in principle all suitable for operation at their nominal point.

---

[2] Effectiveness of the heat exchanger for the reactor system 2Q in table 3.3 is taken from chapter 2.

[3] Flow distribution of the synthesis loop for the reactor system 2Q in table 3.4 is taken from chapter 2.

3.3.2. Off-design performance

In the following section, analysis is performed to determine whether all five design variants are also operable under off-design operation, which would be important for power-to-ammonia operation. Therefore, the load range *i.e.* minimum and maximum NH_3 production capacities of the design variants (system boundary I) and synthesis loop (system boundary II) are investigated by varying one process variable at a time in the process feed stream: $z_{\text{③},RS} \in \{y_{Ar} \mid \dot{m} \mid y_{H_2} : y_{N_2}\}_{\text{③},RS}$.

System boundary I

The minimum and maximum NH_3 productions resulting from the optimisation of the process variables inert gas concentration, feed flow rate and reactants ratio for reactor systems are given in table 3.5. It can be observed for all the reactor systems that maximum NH_3 production is achieved for zero inert gas and that the reactants ratio varies only slightly from the stoichiometric ratio. With an increase in feed flow rate, NH_3 production increases, but conversion of the reactants decreases (see figures B.3 to B.7 in appendix B) due to less space time. On the other hand, at ratios other than stoichiometric ratios of reactants, NH_3 production must decrease; however for NH_3 production maximisation and ratio of reactants, a change in inlet temperatures is observed, therefore NH_3 production also increases; see figures B.8 to B.12 in appendix B. Among the three process variables, controlling inert gas content in the process feed yields the highest NH_3 production, while decreasing H_2-to-N_2 ratio allows one to obtain significantly lower NH_3 production for all the reactor systems. During renewable energy shortages, decreasing the H_2 intake is beneficial, as it consumes more than 90 % of the energy of the power-to-ammonia process.[24] An increase in Ar concentration in the process feed also results in significantly lower NH_3 production, which also means lower H_2 intake. Practically, in real plants of ammonia synthesis, Ar concentration lies within the 0 to 30 mol % range.[6]

By comparing change in the NH_3 production for three process variables, it is observed that change in inert gas (Ar) concentration allows for a change in NH_3 production at both sides of the nominal production rate, *i.e.* minimum and maximum NH_3 production. Therefore, to understand the effect of a change in a process variable on overall steady-state characteristics of the reactor systems, Ar is considered for analysing the results. First the steady-state characteristics for the reactor system 2Q are presented and discussed, and afterwards for the other reactor systems.

Table 3.5.: Minimum and maximum NH_3 production for variation of the process feed: inert gas concentration, flow rate and reactants ratio

RS	Off-Design	Inert gas concentration		Flow rate		Reactants ratio	
		$Y_{Ar,③}$ / mol %	μ_{NH_3} / kg h^{-1}	$\dot{m}_③$ / kg h^{-1}	μ_{NH_3} / kg h^{-1}	$H_{2③}$:$N_{2③}$ / $\frac{mol\ of\ H_2}{mol\ of\ N_2}$	μ_{NH_3} / kg h^{-1}
2Q	MIN	11.62	77.95	527.78	101.53	1.18 : 2.82	38.71
	MAX	0.00	149.50	687.70	122.44	2.998 : 1.002	121.11
HQ	MIN	12.63	75.88	535.26	105.34	1.23 : 2.77	38.63
	MAX	0.00	152.58	696.56	125.67	3.003 : 0.997	123.70
QH	MIN	6.48	117.16	532.30	73.22	1.36 : 2.64	43.35
	MAX	0.00	160.11	661.02	129.09	2.98 : 1.02	129.64
2H-2	MIN	7.30	115.30	565.20	116.84	1.34 : 2.66	43.39
	MAX	0.00	161.87	662.68	131.38	2.99 : 1.01	131.76
2H-3	MIN	9.33	105.25	604.34	124.36	1.31 : 2.69	42.49
	MAX	0.00	163.00	672.42	132.83	2.996 : 1.004	132.40

The load range evaluation of the reactor system 2Q by varying the inert gas concentration in process feed is carried out by stability analysis and presented in figure 3.5. For autothermal operation of a reactor system, it is necessary that the heat generation curve (S-shaped) intersects the heat removal (HE 1) line. For minimum, normal and maximum NH_3 production, the heat generation curves intersect with the heat removal line (HE 1) at three points. For normal and maximum NH_3 production, the optimum is located at the stable higher intersection point. Whereas, for the minimum NH_3 production, the middle intersection is identified. The argon free process feed for determining maximum NH_3 production is satisfied by chapter 2 findings, see figure 2.6. However, the higher argon concentration in the process feed for minimum NH_3 production remains lower, by 1.11 mol %, as reactor system preferred to operate at lower inlet temperature and therefore resulted in much lower NH_3 production. The difference between the two findings of higher argon gas concentration lies due to different objective functions, as chapter 2's primary objective was to determine the operating envelope of process variables, whereas the objective of this chapter is to determine load range of the reactor systems by varying process variables. For argon gas concentration between 11.62 and 12.73 mol %, the reactor system will result in a slightly higher intersection point between the heat generation and heat removal lines. Inert gas concentrations higher than 12.73 mol % will result in less generation of heat than removal of heat, which is not suitable for autothermal operation of the reactor system.

Figure 3.6 reveals the reactants conversion and temperature profiles along the catalysts beds.

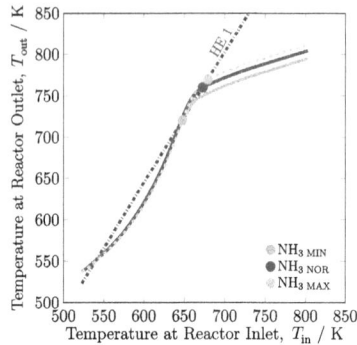

Figure 3.5.: The reactor system 2Q steady-state characteristics for normal (NOR), minimum (MIN) and maximum (MAX) NH_3 production by varying argon concentration.

In figure 3.6a, it can be seen that at a stoichiometric ratio, the reactants H_2 and N_2 conversion overlap each other. For minimum NH_3 production, the reactor system prefers to operate at lower temperatures in all three catalyst beds, whereas for a maximum NH_3 production, it operates at higher temperatures in all catalyst beds, see figure 3.6b. The reactants conversions for maximum NH_3 production remains quite similar to nominal conversions, as with just 5 mol % inert removal, rate of reaction does not vary much.

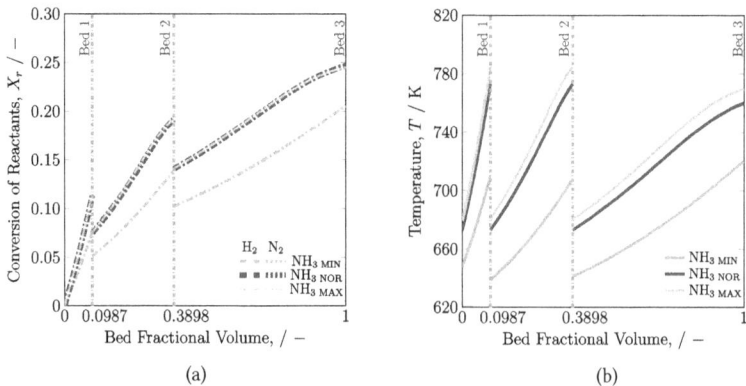

(a)   (b)

Figure 3.6.: The reactor system 2Q (a) reactants conversion profiles and (b) temperature profiles for normal (NOR), minimum (MIN) and maximum (MAX) NH_3 production by varying argon concentration.

The load range results for the reactor system HQ and QH are also analysed in-depth in figures 3.7a and 3.7b. The behaviour of the heat generation curves is similar to the reactor sys-

tem 2Q, however the heat removal lines consist of two elements due to presence of two heat
exchangers: HE 1 and 2 for reactor system HQ, and HE 1 and 3 for the reactor system QH. In
the reactor system HQ, the heat removal line related to HE 2 does not connect directly with
reactor inlet and outlet streams, but it determines inlet temperature of HE 1. Likewise, in reactor system QH, the heat removal line related to HE 3 does not connect with reactor inlet
and outlet streams, it determines inlet temperature of HE 1, see figure 3.1. Furthermore, it can
be seen that the heat removal lines of HE 2 and HE 3 for minimum, normal and maximum
NH_3 production overlap, as process feed inlet temperature and heat exchanger effectiveness
are constant. In contrast, heat removal lines for HE 1 of the reactor systems HQ and QH are
moved in parallel when varying ammonia production, due to variable cold side inlet temperature and constant effectiveness, which is evident from figure 2.10. In comparison to reactor
system QH, reactor system HQ is capable of significantly lowering NH_3 production by using
almost two times higher argon concentration in process feed, see table 3.5. The possible reason
behind this is design; due to quench-based cooling between catalyst beds 2 and 3, reactor system HQ is able to operate at a much lower temperature. On the other hand, reactor system QH
is capable of producing up to 7.5 kg/h more NH_3 from reactor system HQ.

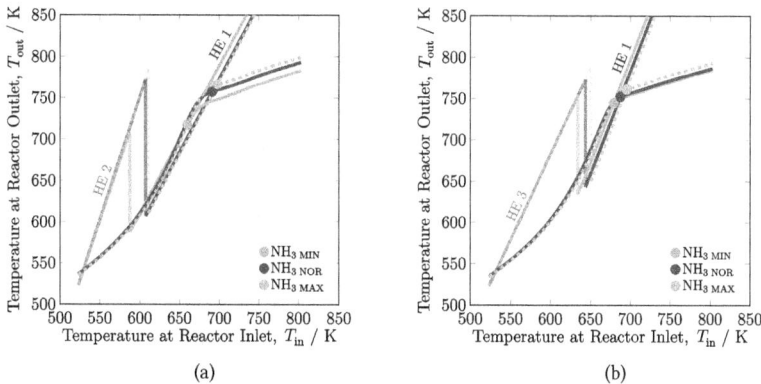

Figure 3.7.: Steady-state characteristics of the reactor systems (a) HQ and (b) QH for normal (NOR),
minimum (MIN) and maximum (MAX) NH_3 production by varying argon concentration.

Finally, the reactor systems with heat exchangers, 2H-2 and 2H-3, are compared to see if omis-
sion of quenching and the sequence of the heat exchangers (in 2H-2, process feed exchanges
heat first in HE 2, and in 2H-3, process feed exchanges heat first in HE 3) has an impact. The
load range of reactor system 2H-2 and 2H-3 by varying the argon gas concentration in pro-

cess feed is given in figures 3.8a and 3.8b. The behaviour of the heat generation curves is quite similar to the other three reactor systems, whereas the heat removal line consists of three elements, HE 1, 2 and 3. From figure 3.8a, it can be seen that the process feed enters first in HE 2 at 523 K and afterwards enter HE 3 and 1 at the desired operational temperatures. On the other hand, from figure 3.8b it can be seen that feed first enters in HE 3 at 523 K and afterwards passes through HE 2 and 1 at desired operational temperatures. Furthermore, from the two figures 3.8a and 3.8b, with the help of the distance between minimum and maximum NH_3 production operational points, the production flexibility comparison between the two can also be made. In figure 3.8a, it can be seen that all three operational points are quite close to each other, whereas for reactor system 2H-3 (see figure 3.8b), the minimum NH_3 production operational point is at a bit of a distance, therefore it should provide more provisions for operation at lower NH_3 production and likewise more argon concentration in the process feed, which is also evident from table 3.5.

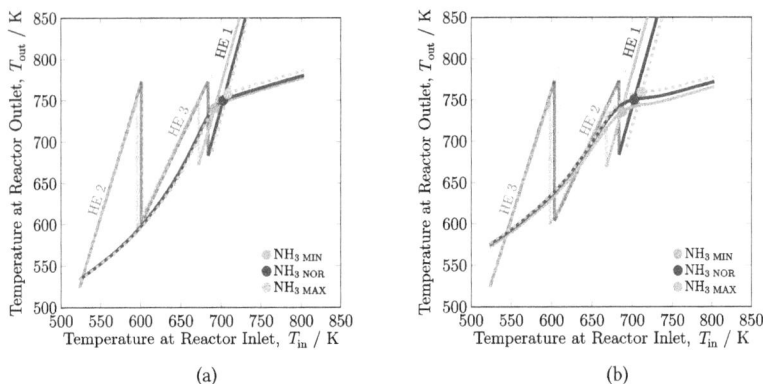

(a)   (b)

Figure 3.8.: Steady-state characteristics of the reactor systems (a) 2H-2 and (b) 2H-3 for normal (NOR), minimum (MIN) and maximum (MAX) NH_3 production by varying argon concentration.

From the results of the minimum and maximum NH_3 production from variations of process variables, table 3.5 and from figures 3.5 to 3.8b and figures B.1 to B.12 in appendix B, it is concluded that H_2-to-N_2 ratio provides the maximum variation in NH_3 production. Furthermore, the maximum operational temperature condition 803 K was approached in bed 1 of reactor systems 2Q, HQ, 2H-2 and 2H-3 for NH_3 production minimisation by lowering mass feed flow rate only, see figures B.3b, B.4b, B.6b and B.7b (appendix B).

System boundary II

The graphical summary of the flexibility analysis of NH_3 production, H_2 intake, operational, recycle load and recycle to feed ratio within the synthesis loop performed for all design variants with regards to the process variables is presented in figure 3.9. It can be concluded that NH_3 production strongly correlates with H_2 intake, *i.e.* with increase in NH_3 production, H_2 intake also increases. Recycle load and recycle to feed ratio are inversely proportional to NH_3 production by varying inert gas concentration and reactants ratio. Also, NH_3 production is inversely proportional to inert gas concentration, see figure 3.9a. Furthermore, it can be seen that NH_3 production does not significantly increase with varying feed flow rate (see figure 3.9b) and reactants ratio (see figure 3.9c), however with their variations, NH_3 production can be reduced. However, with inert gas concentration variations, the significant NH_3 production load can be increased and decreased. In contrast to the other reactor system, QH provides significantly lower ammonia production by decreasing feed flow rate, however it increases recycle to feed ratio, as with lower feed flow rate it also reduces inlet temperature and thus it reduces reactants conversion and enhances recycle load, see figures B.5a and B.5b in appendix B.

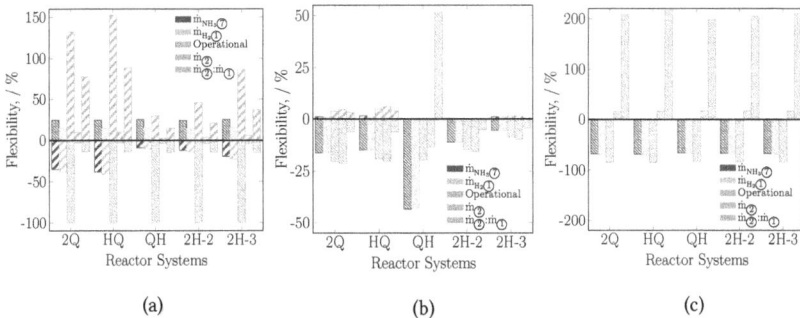

Figure 3.9.: Overall NH_3 production, H_2 intake, operational, recycle load and recycle to feed ratio flexibilities for manipulable process variables: (a) inert gas, (b) feed flow rate and (c) reactants ratio in the synthesis loop for the design variants.

After comparing results of optimisation for minimum and maximum NH_3 production for three process variables from figures 3.9a to 3.9c, it is concluded that H_2-to-N_2 ratio has the most potential to reduce H_2 consumption by up to 70 %. The reduction of NH_3 through reactants ratio also reduces H_2 intake significantly by 70 %, the reduction in H_2 intake is quite useful for a renewable energy outage period. Whereas, inert gas (Ar) free synthesis loop can increase NH_3

production by 15 % in all design variants, but in reality, Ar free fresh feed is difficult to maintain and therefore for the maximum NH_3 production, Ar gas concentration in the synthesis loop needs to be maintained as low as possible.

3.4. Conclusions

This chapter presented a systematic analysis of five different autothermal ammonia synthesis reactor systems with regard to NH_3 production minimisation and maximisation. From the results, it can be concluded that all five autothermic reactor systems are viable for power-to-ammonia process, as they can be operated for a wide range of ammonia production flexibilities. Among the five design variants for minimum NH_3 production by increasing Ar gas concentration in the synthesis loop, 2Q and HQ seem more significant. On the other hand, QH provides only significantly minimum NH_3 production with variation in mass feed flow rate. Whereas for reactants ratio variation, all design variants behave quite similarly and the same is the case with a decrease in the inert gas concentration in the synthesis loop.

4. Performance comparison of ammonia synthesis loops

As mentioned earlier, in addition to the ammonia synthesis reactor system, ammonia production is influenced by the ammonia separator, as well as by the recycle loop. Therefore in this chapter, the model-based flexibility analysis is extended with the inclusion of a separator model. Furthermore, different ammonia synthesis loop configurations are considered, such as ammonia separator after the reactor system and before the reactor system. Thereafter, the synthesis loops are compared with regard to power-to-ammonia requirements, such as flexible ammonia production and hydrogen intake. In addition, from the results of the chapters 2 and 3, it is observed that by changing process variables from the nominal value, ammonia production also changes. Some variables have positive or negative correlation with ammonia production. For example, with an increase in pressure, ammonia production increases, whereas with an increase in Ar content in process feed, ammonia production decreases. Therefore, in this chapter, multi-variable optimisation is also applied for enhancing the ammonia production load range. This chapter is organised as follows: first, literature review regarding ammonia synthesis loop configurations is made. Afterwards, a model of ammonia separator and multi-variable optimisation strategy is presented, followed by results and discussion.

4.1. Ammonia synthesis loops

The ammonia synthesis loops are classified with respect to the location of the fresh feed entrance and the ammonia separator, see figure 4.1. Therefore, various configuration possibilities exist; for example, ammonia separator after (loop I) and before (loop II) the reactor system, see figures 4.1a and 4.1b, respectively. In loop I, fresh feed is mixed with recycled reactants before entering the reactor system and in loop II, fresh feed is mixed with product gas just before entering the separator.[6,32,35] In addition, a combination of both loops I and II can also exist. For instance, two ammonia separators could be installed in the synthesis loop, one before and one after the reactor system.[6,59] In all of these configurations, operational pressures and temperatures are regulated by compressors and a trail of heat exchangers. For example, the loops I and II process flow diagrams can be seen in figure 2.1 (chapter 2) and figure C.1 (appendix C), respectively.

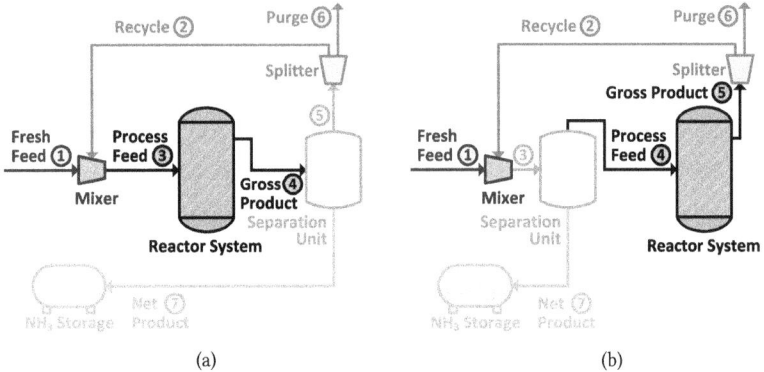

(a)   (b)

Figure 4.1.: Haber-Bosch ammonia synthesis loop configurations: (a) with ammonia separation unit after the reactor system (loop I) and (b) with ammonia separation unit before the reactor system (loop II).

Fresh feed quality plays a decisive role in selecting the synthesis loop configurations. Loop I configuration is preferred for a pure and dry fresh feed. Whereas, loop II configuration is preferred for an impure and wet fresh feed, so poisonous gases like CO_2 and H_2O coming along with the reactants H_2 and N_2 are removed from the stream by their dissolution in liquid ammonia, before they could harm the catalyst in the reactor system. Both synthesis loops require purging of inert gas(es), as they are not reacting and therefore accumulate in the synthesis loop.[6] With a one-stage ammonia separation in the synthesis loop, up to *ca.* 97 mol % NH_3 purity is achievable.[32,35,60] To achieve higher purity, a two or more staged separation unit is required.[51,61,62]

With regard to capital and operational cost, loop I most likely is more economical than loop II, as the maximum amount of NH_3 is separated after the reactor system, which means in the recycle loop, a low flow rate and low ammonia concentrations need to be handled, and thus less load is exerted on the recycle compressor. Whereas, in loop II for the same condensing conditions, a higher NH_3 content exists along with the reactant stream and enter in the reactor system and therefore results in less reactants conversion. Therefore, loop I requires small sizes of unit operations: reactor, separator, heat exchangers and compressors, and also less intake of reactants, this all leads to less operational energy requirements.[6,32] For power-to-ammonia processes, synthesis loop I is preferable, as clean and dry reactants H_2 and N_2 are obtainable (see figure 1.2).[4,23,24] Economy and efficiency are two aspects of selecting the process, whereas power-to-ammonia depends on the intermittent renewable energy and therefore also requires flexible operation and ammonia production. So far, comparisons between ammonia synthesis loops have been conducted with regard to economy[6,32] and process control[35,51,60–62] only. This gives the clear motivation in this section to inquire suitable ammonia synthesis loop configurations for the power-to-ammonia process with regard to the flexibility in ammonia production.

The focus of this chapter is to compare synthesis loops I and II with regard to flexible ammonia production. Therefore, the unit operations which require heat balance only are neglected, *i.e.* heat exchangers, coolers and compressors with inter-stage cooling; their involvement will be of interest for acquiring synthesis loop energy requirements and it is out of scope for this chapter. However, for the reactor system (heat exchanger (equation 3.1b), catalyst beds (equation 2.3) and mixers (equation 2.6)) energy balance is considered, as reactor system is autothermal and its operation is strongly dependent on balance between heat production and removal. Whereas, for unit operations other than reactor systems, it is assumed that they rely on utility usage for heating and/or cooling purposes, and the supply of the utilities is adjustable accordingly. Design of the unit operations also does not hinder in load variation of ammonia production. These assumptions are made in accordance of Friedrichsen[63] findings, and valid for *ca.* 10 to 170 % load range of ammonia production. For comparison between synthesis loops, first mathematical modelling and the assumptions regarding the involved operations in the synthesis loops are discussed. Afterwards, possible optimisation strategies for the reactor systems of the synthesis loops I and II are defined. The reactor systems in both synthesis loops differ, as both loops have different process feed composition and flow rate, thus resulting in different energy requirements.[32] Therefore, for the design-based performance comparison between the synthesis loops, reactor systems needed to be at their optimum. As the scope of this work is to do flexibility analysis of the loops, therefore without going into design details of the reactor systems, overall reactor systems volume is inquired similarly to chapter 2. Eventually, the methodologies for enhancing load range of ammonia production, *i.e.* minimum and maximum net product for the synthesis loops are discussed, so the flexibility analysis can be performed.

4.2. Methodology

For performing the comparative flexibility analysis between the synthesis loops shown in figure 4.1, the model is kept compact and simplified by considering various assumptions. Other than the ammonia separation unit, the model for the reactor system and the ammonia synthesis loop is taken from chapter 2. However, the synthesis loop model is reintroduced in more generalised terms, so it can be fitted to both of the synthesis loops. Therefore, first the mathematical model of the separation unit along with suitable assumptions is presented. Afterwards, the multi-variable optimisation problem formulation for different scenarios, such as design and off-design, is made. At the end, implementation of the optimisation problems is discussed.

4.2.1. Mathematical model

As the target of this work is to complete the flexibility analysis of the synthesis loops by acquiring minimum and maximum ammonia production limitations, along with fresh hydrogen intake, therefore for unit operations other than the reactor system, only the material balance is implemented. The mathematical model of the separation unit and the overall ammonia synthesis loop is presented as follows.

Separation unit

For achieving a product purity of 99.99 mol % NH_3, a two-stage ammonia separation unit (SU) is considered. $S \in \{1, 2\}$ refers to the two separators (stages). Besides separators, the SU also contains a recycle loop for recycling back vapours from S2 to S1, additionally a compressor and a cooler are also installed in the recycle stream. Thus, components of both streams can mix at same conditions, see figure 4.2a. It is assumed that the separators are operated at constant pressure P_S and temperature T_S, so called PT-flash separators. Furthermore, instantaneous separation is considered, as gas phase hold-up time within a PT-flash separator is assumed to be negligible. The outlet composition of the gaseous top stream $Y_{S,c_{T_{out}}}$, the liquid bottom stream $Y_{S,c_{B_{out}}}$ and the ratios between the two streams $Y_{S,c_{T_{out}}} : Y_{S,c_{B_{out}}}$ for each PT-flash separator are determined. As low ammonia storage capacity (below 1500 t)[6] is considered, therefore the two separators of the synthesis loops will be preferably operated at ambient temperature of *ca.* 298 K. However, the two separators will operate at different P_S: the first separator (S1) will operate at the synthesis loop operational pressure, *i.e.* between (150 to 220 bar), whereas the second separator (S2) will operate at much lower pressure, preferably just above the dew point of ammonia *ca.* 12 bar, so the flashing of $Y_{S1,c_{B_{out}}}$ can take place in S2.

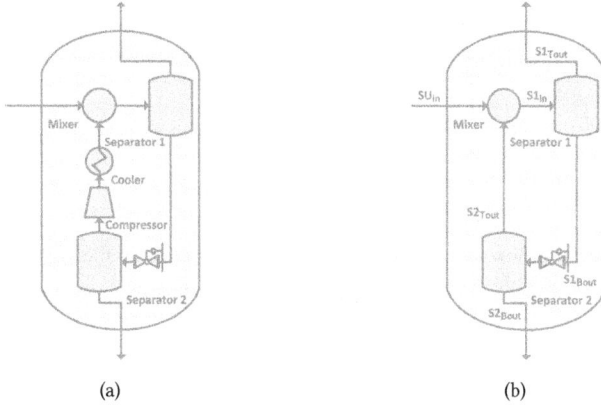

(a)   (b)

Figure 4.2.: Two-stage ammonia separation unit (a) configuration and (b) lumped model.

The overall material balance for the separators $S \in \{1,2\}$ of the synthesis loops $L \in \{I, II\}$ for components $c \in \{N_2, H_2, NH_3, Ar\}$ is given as follows:

$$\dot{n}_{S_{in},L} \, y_{c,S_{in},L} = \dot{n}_{S_{T_{out}},L} \, y_{c,S_{T_{out}},L} + \dot{n}_{S_{B_{out}},L} \, y_{c,S_{B_{out}},L} \tag{4.1}$$

$$y_{c,S_{T_{out}},L} = K_{c,S,L} y_{c,S_{B_{out}},L} \tag{4.2}$$

where, $\dot{n}$ is the molar flow rate, y_c is molar fraction and K_c is vapour-liquid equilibrium constant for the components $c \in \{N_2, H_2, NH_3, Ar\}$. Subscripts: S_{in}, $S_{T_{out}}$ and $S_{B_{out}}$ presents the inlet, top and bottom streams of the separators $S \in \{1,2\}$, see figure 4.2b. For the above mentioned operating range of two separators, temperatures of the N_2, H_2 and Ar are above the critical temperatures, whereas temperature of NH_3 is below critical temperature. Therefore, for determining K_c of the components N_2, H_2 and Ar, Henry's law (equation A.37, appendix A) will be used and for NH_3, Raoult's law (equation A.38, appendix A) will be used.[54] The supporting equations A.37 to A.42 for the separator are provided in appendix A. In addition, it is pertinent to mention that all molar flow rates $\dot{n}$ and components molar composition y_c of all streams $\textcircled{S}$ in the loops are converted to mass flow rate $\dot{m}$ and mass fractions x_c, as with involvement of reaction for overall material balance, it is more efficient to maintain a check on mass balance than mole balance.

For the mixer between the separators, the mixing of the components $c \in \{N_2, H_2, NH_3, Ar\}$ among the inlet stream of separation unit (SU) and top stream of the separator 2 (S2) is considered to take place at same pressure and temperature conditions (see figure 4.2), and presented as follows:

$$\dot{m}_{c,S1_{in},L} = \dot{m}_{c,S2_{T_{out}},L} + \dot{m}_{c,SU_{in},L} \tag{4.3}$$

where, $\dot{m}$ is the mass flow rate and the subscript $L \in \{I, II\}$ refers to the synthesis loop. The overall material balance for the separation unit is given as follows:

$$\dot{m}_{c,SU_{in},L} = \dot{m}_{c,S1_{T,out},L} + \dot{m}_{c,S2_{B,out},L} \tag{4.4}$$

In the context of figure 4.1, for the separation unit of the loops I and II, the inlet and outlet streams are expressed as $\dot{m}_{c,SU_{in},I} = \dot{m}_{c,\textcircled{4},I}$, $\dot{m}_{c,SU_{in},II} = \dot{m}_{c,\textcircled{3},II}$, $\dot{m}_{c,S_{T,out},I} = \dot{m}_{c,\textcircled{5},I}$, $\dot{m}_{c,S1_{T,out},II} = \dot{m}_{c,\textcircled{4},II}$ and $\dot{m}_{c,S2_{B,out},L} = \dot{m}_{c,\textcircled{7},L}$.

Ammonia synthesis loop

To evaluate the overall performance of the ammonia synthesis loops, all unit operations models are required to be combined. The recycle stream influences the composition within the synthesis loop streams, see figure 4.1. The material balance of the synthesis loops $L \in \{I, II\}$ with regard to components $c \in \{N_2, H_2, Ar\}$ is calculable by equations A.32, A.33 and A.34. As these equations were initially developed for only loop I, therefore for the loop II stream $\textcircled{4}$ need to be replaced with stream $\textcircled{5}$. For applicability of equations on both the synthesis loops, equations are transformed again as follows:

$$\dot{m}_{N_2,\textcircled{3},L} - \dot{m}_{N_2,\textcircled{P},L} + p_L \, \dot{m}_{N_2,\textcircled{P},L} - x_{N_2,\textcircled{1},L} \, \dot{m}_{\textcircled{1},L} = 0 \tag{4.5}$$

$$\dot{m}_{Ar,\textcircled{3},L} - \dot{m}_{Ar,\textcircled{P},L} + p_L \, \dot{m}_{Ar,\textcircled{P},L} - 0.0285 x_{N_2,\textcircled{1},L} \, \dot{m}_{\textcircled{1},L} = 0 \tag{4.6}$$

$$\dot{m}_{H_2,\text{③},L} - \dot{m}_{H_2,\text{ⓟ},L} + p_L \, \dot{m}_{H_2,\text{ⓟ},L} - (1 - 1.0285 x_{N_2,\text{①},L}) \, \dot{m}_{\text{①},L} = 0 \qquad (4.7)$$

where, gross product stream ⓟ ∈ {4 or 5} with regard to the synthesis loops $L \in \{I, II\}$, respectively, see figure 4.1. By solving equations 4.5 to 4.7 simultaneously, the purge ratio p_L, the fresh feed flow rate $\dot{m}_{\text{①},L}$ and the N_2 mass fraction $x_{N_2,\text{①},L}$ in fresh feed become known for the respective synthesis loop. Afterwards, by implementation of equations A.30 and A.31, Ar and H_2 mass fractions are evaluated. Furthermore, material balance equations for the mixer (equations A.24 and A.25) and the splitter (equations A.19 to A.22) remained as is, and can be applied to calculate the mass flow rates of purge stream $\dot{m}_{\text{⑥}}$ and recycle stream $\dot{m}_{\text{②}}$.

4.2.2. Problem formulation

In this section, to perform flexibility analysis of the ammonia synthesis loops, a two step strategy has been adopted. First, the design performance problem for determining the minimum reactor volume for a given production rate is formulated and afterwards the off-design performance analysis problem for determining minimum and maximum ammonia production capacities is formulated. For the fair performance comparison between two loops, common design and operational criteria have been considered.

Design performance

To determine the optimum volume of the catalyst beds of the reactor systems and to compare the design performance among both synthesis loops, the parameters and process variables are kept constant. The fresh feed (stream ①) composition and the operational conditions of reactor system are taken from table 2.1. However, unlike in chapter 2, the composition of process feed ⓕ of the synthesis loops is set free, as now, in the synthesis loop, the separator model has been taken into account. Additionally, instead of hit and trial method for finding the optimum volume of the reactor system, a mathematical optimisation problem is defined.

First, the objective function of the optimisation problem for minimum volume of the three-bed reactor systems is formulated for the synthesis loops. Like in chapter 2, the reactor systems are also designed for the production of 120 kg/h of NH_3, which is capable of producing approximately 50 MWh/day of energy via a 29 % efficient internal combustion engine.[21] Furthermore, the reactor system is preferably designed for 90 % of the equilibrium conversion of reactants, which helps to avoid the infinite amount of reactor space requirement for accomplishing equilibrium conversion.[37] For determining the minimum volume of the reactor systems of the loops I and II, the equality and non-equality constraints are implemented, in addition to the equations related to the reactor system (equations 2.2 to 2.6), separation unit (equations 4.1 to 4.4) and ammonia synthesis loop (equations 4.5 to 4.7, A.22 and A.24), as follows:

$$\underset{z_L}{\text{minimise }} V_L = \sum_{b=1}^{3} V_{b_L} \tag{4.8}$$

subject to

Equations 2.2 to 2.6 (reactor system).

Equations 4.1 to 4.4 (separation unit).

Equations 4.5 to 4.7 (synthesis loop).

Equations A.22 and A.24 (synthesis loop).

$$673\,\mathrm{K} = T_{b_{\mathrm{in}},RS} < T_{b_{\mathrm{out}},RS} = \begin{cases} 773\,\mathrm{K}; \text{ for } b \in \{1,2\} & \text{(4.9a)} \\ 0.90\,T_{EQ,RS}; \text{ for } b = 3 & \text{(4.9b)} \end{cases}$$

where, the overall reactor system volume V_L is composed of the sum of the volumes V_{b_L} of the individual catalyst beds $b_L \in \{1, 2, 3\}$. The optimisation variables are $z_L \in \{\dot{m}_{\textcircled{F},L}, Y_{c,\textcircled{3},L}, V_{b_L}\}$. For the synthesis loop II, $\dot{m}_{\textcircled{3},\mathrm{II}}$ is a additional optimisation variable. For synthesis loops $L \in \{\mathrm{I}, \mathrm{II}\}$, process feed $\textcircled{F}_L \in \{3, 4\}$, the composition of the components $Y_{c,\textcircled{3},L} \in \{\mathrm{N}_2, \mathrm{H}_2, \mathrm{NH}_3, \mathrm{Ar}\}$ and the mass flow rate of process feed $\dot{m}_{\textcircled{F},L}$. The mass flow rate of the process feed is further subdivided into four streams, see figure 3.1a of the direct cooling by quenching reactor system (2Q). As earlier it is stated that the quench stream between heat exchanger and catalyst bed 1 is not in operation, therefore $\dot{m}_{\textcircled{F},L} \in \{\dot{m}_{b1_L}, \dot{m}_{q2_L}, \dot{m}_{q3_L}\}$. Like chapter 2, the inlet $T_{b_{\mathrm{in}}}$ and the outlet $T_{b_{\mathrm{out}}}$ temperature of beds $b \in \{1, 2\}$ are maintained to 673 and 773 K, respectively. Whereas for bed 3 the inlet temperature is maintained to 673 K and the outlet to 90 % of an equilibrium temperature $T_{EQ,L}$. For the synthesis reactor design, H_2-to-N_2 molar ratio is maintained at 3 : 1, and $\sum_{c=1}^{4} Y_{c,\textcircled{3},L} = 1$ within the synthesis loop is assured. The inert gas concentration is controlled in the synthesis loops by purging. The purge ratio is fixed ($p_L = 0.02$), it is an acceptable value for maintaining Ar gas concentration around 5 mol %, see table 3.4.

The boundary limits of the process variables and the design parameters for the design performance analysis are stated in table 4.1. For the design performance, the boundaries can be set free, however for avoiding extra computation time, the boundary conditions were implemented and narrowed down with regard to the chapter 2 optimum results shown in table 2.2.

Table 4.1.: Boundary limits of the process variables and the design parameters for the design performance analysis of the synthesis loops

Process variables	Lower bound	Upper bound
$\dot{m}_{b1,L}$ / $kg\,h^{-1}$	200	400
$\dot{m}_{q2,L}$ / $kg\,h^{-1}$	100	200
$\dot{m}_{q3,L}$ / $kg\,h^{-1}$	150	250
$\dot{m}_{\textcircled{3},\text{II}}$ / $kg\,h^{-1}$	500	1000
$Y_{H_2,\textcircled{3},L}$ / mol %	0	100
$Y_{N_2,\textcircled{3},L}$ / mol %	0	100
$Y_{NH_3,\textcircled{3},L}$ / mol %	0	100
$Y_{Ar,\textcircled{3},L}$ / mol %	0	15

Design parameters	Lower bound	Upper bound
$V_{b1,L}$ / m^3	0.0050	0.0100
$V_{b2,L}$ / m^3	0.0100	0.0300
$V_{b3,L}$ / m^3	0.0300	0.0600

Off-design performance

For the off-design performance analysis, the minimum and maximum limitations of ammonia production of the synthesis loops are evaluated for the case of autothermal operation of the reactor system. Autothermal operation requires one to take into account the heat management between the heat of production and removal. Therefore, the stability range of the reactor systems for optimisation needs to be defined. The stability of the reactor system is assessed by using the van Heerden steady state stability approach.[39] For implementation of off-design optimisation, an approach similar to chapter 3 is adopted. In addition to equations 2.2 to 2.6, 4.1 to 4.4, 4.5 to 4.7, A.22 and A.24, equality and inequality constraints (equations 3.1b, 3.3 and 3.4) are implemented. Where, in equation 3.1b, effectiveness ε of the heat exchanger is evaluated for the optimum temperatures calculated during the design performance analysis. The optimisation problem for minimum and maximum ammonia production is formulated as follows:

$$\text{minimise} \mid \underset{z_L}{\text{maximise}} \ \dot{m}_{\textcircled{7},L} \tag{4.10}$$

subject to

Equations 2.2 to 2.6 (reactor system).

Equations 3.1b, 3.3 and 3.4 (reactor system).

Equations 4.1 to 4.4 (separation unit).

Equations 4.5 to 4.7 (synthesis loop).

Equations A.22 and A.24 (synthesis loop).

$$0 < p_L \le 0.10 \tag{4.11}$$

where $\dot{m}_{\textcircled{7},L}$ is the product from the synthesis loops $L \in \{I, II\}$ and $z_L \in \{P, T_{\textcircled{F},L}, T_{in_L}, \dot{m}_{\textcircled{F},L}, Y_{c,\textcircled{3},L}\}$, in addition, for the synthesis loop II $\dot{m}_{\textcircled{3},II}$ is also considered as an optimised variable. For the synthesis loops $L \in \{I, II\}$, the process feed $\textcircled{F}_L \in \{3, 4\}$ are presented, respectively. Composition of the components $Y_{c,\textcircled{3},L} \in \{N_2, H_2, NH_3, Ar\}$ and mass flow rate of process feed $\dot{m}_{\textcircled{F},L}$ is subdivided as follows, $\dot{m}_{\textcircled{F},L} \in \{\dot{m}_{b1_L}, \dot{m}_{q2_L}, \dot{m}_{q3_L}\}$. In addition, $\sum_{c=1}^{4} Y_{c,\textcircled{3},L} = 1$ is also assured. Furthermore, change in $Y_{N_2,\textcircled{3},L}$ and $Y_{H_2,\textcircled{3},L}$ concentration is inversely proportional to each other, $i.e.$ for net product $\dot{m}_{\textcircled{7},L}$ minimisation, $Y_{H_2,\textcircled{3},L} < Y_{N_2,\textcircled{3},L}$ and for net product $\dot{m}_{\textcircled{7},L}$ maximisation, $Y_{H_2,\textcircled{3},L} > Y_{N_2,\textcircled{3},L}$ is also implemented. Furthermore, the change in process feed flow rate is taken proportionally to the changes in the quench flow rates and the feed rate of bed 1.

The boundary limits of the process variables for the off-design performance analysis are stated in table 4.2.The inert gas concentration, temperature range and lower pressure limits are the usual operating envelopes taken from literature (also mentioned earlier in chapter 2).[6] Process feed flow rate ranges between 55 and 115 % of load, as this limit is imposed by centrifugal compressor.[64] The upper pressure limit is considered to be 10 % higher than normal operation pressure; this limit is imposed by vessel design considerations.[58]

Table 4.2.: Boundary limits of the process variables for the off-design performance analysis of the synthesis loops

Process variables	Lower bound	Upper bound
P_L / bar	150	220
$T_{\textcircled{E},L}$ / K	450	550
$T_{in,L}$ / K	623	773
$\dot{m}_{\textcircled{3},L}$ / kg h^{-1}	0.55 $\dot{m}_{\textcircled{3},NOR,L}$	1.15 $\dot{m}_{\textcircled{3},NOR,L}$
$\dot{m}_{\textcircled{4},II}$ / kg h^{-1}	0.55 $\dot{m}_{\textcircled{4},NOR,II}$	1.15 $\dot{m}_{\textcircled{4},NOR,II}$
$Y_{H_2,\textcircled{3},L}$ / mol %	0	100
$Y_{N_2,\textcircled{3},L}$ / mol %	0	100
$Y_{NH_3,\textcircled{3},L}$ / mol %	0	100
$Y_{Ar,\textcircled{3},L}$ / mol %	0	15

4.2.3. Problem implementation

The optimisation problems are implemented in MATLAB. Fmincon is called for the optimisation problem, and for solving the system of equations of the catalyst beds (equations 2.2 and 2.3) a built-in ODE 45 solver is used. This continues iteratively until the constraint conditions are met and no further minimisation or maximisation is possible. The optimisation problem was initialised with the use of multiple starting points, and done repeatedly until the objective function's further reduction or increment is no longer possible.

It is foremost to mention that for the synthesis loop I, the optimiser initialises the design performance problem from the reactor system, and passes components compositions and stream flow rates to the ammonia separator, see figure 4.3a. Whereas, for the synthesis loop II, the optimiser begins the design performance problem from the ammonia separator before the reactor system, see figure 4.3b. The difference in optimisation strategy lies between the two loops, due to the position of stream ③, and initialising data related to this stream is provided. A similar sequential strategy has been adapted by the optimiser for the off-design performance, see figure 4.4.

4.3. Results and discussions

The synthesis loops' design performance results for minimum volume of the reactor systems are presented along with the optimum process variable values. Afterwards, the off-design performance results for minimum and maximum ammonia production by varying process variables together are presented. Finally, for the off-design performance analysis, flexibilities

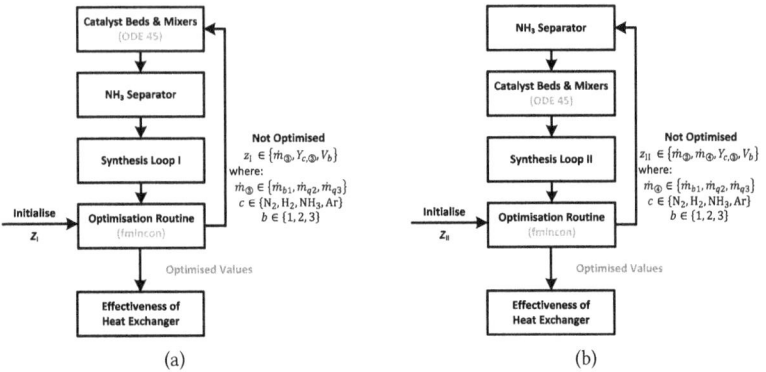

(a)   (b)

Figure 4.3.: Flow chart of the design performance optimisation strategy for (a) ammonia separator unit after the reactor system (loop I) and (b) ammonia separator unit before the reactor system (loop II).

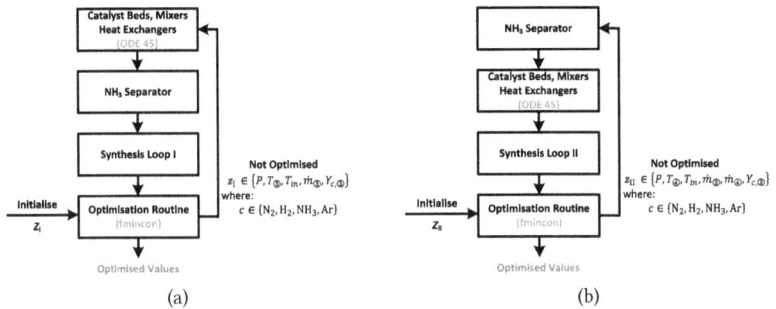

(a)   (b)

Figure 4.4.: Flow chart of the off-design performance optimisation strategy for (a) ammonia separator unit after the reactor system (loop I) and (b) ammonia separator unit before the reactor system (loop II).

of net product, H_2 intake, recycle load, recycle-to-feed ratio and purge fraction of synthesis loops are presented and discussed.

4.3.1. Design performance

In this section, results obtained for minimum volume of the reactor systems of the loops I and II for the same initial conditions are presented. In addition, the design optima of process feed distribution within catalyst beds, process feed compositions and volume of catalyst beds are presented and discussed. Afterwards, temperature-reactants conversion $\overline{TX}$ trajectories are illustrated. Finally, effectiveness of the heat exchangers in the reactor systems and the flow rate distributions in the syntheses loops are analysed.

The minimum volume of the reactor systems from the optimum process variables feed flow rate of bed 1, quench stream flow rate, stream ③ composition and flow rate are given in table 4.3. It can be seen that the volume of the catalyst beds of the synthesis loop I are smaller than the volume of the catalysts beds of the synthesis loop II. The main reason is that the ammonia content in the process feed of synthesis loop II is *ca.* 32 % higher than that of loop I. Therefore, as a consequence of higher ammonia content in the stream entering the reactor system, for the same product rate, a higher process feed rate is required and thus as a result, catalyst beds of larger volumes are required. Smaller reactor volume and process feed flow rate make the synthesis loop I more economically viable than the synthesis loop II. As capital and operational costs are proportional to the size of the reactors and process feed flow rate, *e.g.* small size reactors cost is less also requires less catalyst, low flow rate in synthesis loop means less reactants intake, lower load on synthesis loop compressors and lower utility requirements in heat exchangers.[6]

The trajectories of temperature and reactants conversion of the synthesis loops I and II are given in figure 4.5a and 4.5b, respectively. It is evident that the reactor systems of both synthesis loops are operated for the maximum possible temperature and reactants conversion, *i.e.* 90 % of equilibrium (EQ). Reactor systems of both loops resulted in almost equal conversion, the reactor system of loop I resulted in just 0.14 % more reactants conversion. This is despite the fact that the process feed of the reactor system of loop II contains almost 32 % more NH_3 content and also handled *ca.* 5 % more flow rate. Basically, for a similar resulting output of reactants conversion, the loop II requires *ca.* 23 % larger reactor volume from loop I,

Table 4.3.: Optimal process variables and parameters (highlighted in grey colour) for the synthesis loops and resulting minimum volume (enclosed in rectangle) of the reactor systems

Loop	Volume of catalyst beds			Reactor system volume
	V_{b1} / m^3	V_{b2} / m^3	V_{b3} / m^3	V / m^3
I	0.0069	0.0208	0.0448	0.0725
II	0.0092	0.0272	0.0526	0.0890

Loop	Flow rate of quench streams			
	$\dot{m}_{q1}$ / kg h^{-1}	$\dot{m}_{q2}$ / kg h^{-1}	$\dot{m}_{q3}$ / kg h^{-1}	$\dot{m}_{q\,Total}$ / kg h^{-1}
I	-	166.74	182.28	349.02
II	-	176.29	186.66	362.95

Loop	Flow rate within catalyst beds			
	$\dot{m}_{b1}$ / kg h^{-1}	$\dot{m}_{b2}$ / kg h^{-1}	$\dot{m}_{b3}$ / kg h^{-1}	$\dot{m}_{\textcircled{F}}$ / kg h^{-1}
I	326.63	493.36	675.64	675.64
II	347.85	524.14	710.80	710.80

Loop	Stream ③				
	Y_{H_2} / mol %	Y_{N_2} / mol %	Y_{NH_3} / mol %	Y_{Ar} / mol %	$\dot{m}_{\textcircled{3}}$ / kg h^{-1}
I	67.79	22.60	3.68	5.93	675.64
II	60.56	20.19	14.00	5.25	830.81

Loop	Stream ⓕ composition			
	Y_{H_2} / mol %	Y_{N_2} / mol %	Y_{NH_3} / mol %	Y_{Ar} / mol %
II	67.01	22.34	4.84	5.81

see table 4.3. The effectivenesses of the heat exchangers of the respective synthesis loops' reactor system are calculated by using equation 3.1b, and the values are ε_L = {0.6296, 0.6462}. As the outlet temperature of bed 3 for loop II reactor system is slightly higher than that of loop I (see figure 4.5), it therefore resulted in a higher effectiveness value.

The fresh feed, purge and recycle loads for the synthesis loops I and II are given in table 4.4. It can be seen that for the synthesis loop II, the recycle and purge loads are significantly higher (*ca.* 30 %) than the values of the synthesis loop I. Therefore, the operational expenses of synthesis loop II will be higher compared to the loop I, as recycle compressor and purge recovery units of higher capacities are required.

Table 4.4.: Design performance: flow distribution in the synthesis loops

Loop	Fresh Feed $\dot{m}_{\textcircled{1}}$ / kg h^{-1}	Purge $\dot{m}_{\textcircled{6}}$ / kg h^{-1}	Recycle $\dot{m}_{\textcircled{2}}$ / kg h^{-1}
I	131.11	11.11	544.53
II	134.22	14.22	696.59

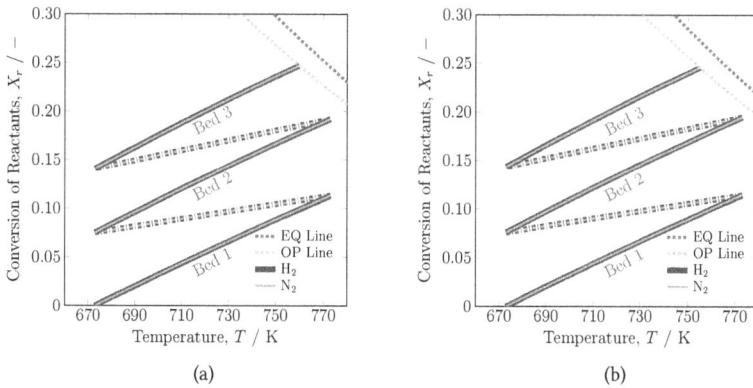

Figure 4.5.: Temperature-reactants conversion $\overline{TX}$ trajectories of reactor system along with equilibrium (EQ) and operational (OP) lines for synthesis loops: (a) the ammonia separator after reactor system (loop I) and (b) the ammonia separator before reactor system (loop II).

4.3.2. Off-design performance

In the following section, the load ranges of the ammonia synthesis loops I and II are presented. For loop I, eight process variables are varied together to identify the widest span of ammonia production; whereas for loop II, nine process variables are varied. To evaluate the load range of the synthesis loops having an autothermal reactor system, the steady-state stability analysis performance of the reactor system is necessary and presented in figure 4.6. Finally, the synthesis loops are analysed and compared with regard to percentage yield and flexibility of NH_3 production, along with H_2 intake, recycle and purge loads.

In table 4.5, it can be seen that for the operational pressure and process feed temperature, the foxed higher boundary limits are reached for minimum and maximum production for both loops. For mass flow rate of process feed, lower and upper boundaries for minimum and maximum ammonia production, respectively, are reached by synthesis loop I only. For minimisation and maximisation of ammonia production in the synthesis loops, the reactor system is preferably operated far away from stoichiometric ratios of reactants. In chapter 2, it was observed that ammonia production decreases for reactants supply at rates other than stoichiometric ratio. Here, however for maximum ammonia production, along with reactants ratio, other process variables also change, e.g. process feed temperature, and operation pressure increase, whereas inert gas concentration in the synthesis loop decreases. Similarly, with increase in process feed flow rate only, it was observed in chapter 2 that ammonia production decreases, as reactants do not have enough time for reaction. But here, other process variables also change, as pressure and inlet temperature increase supporting an increase in process feed intake to maximise ammonia production, see table 4.5.

Table 4.5.: Results of the minimisation and maximisation of NH_3 production of synthesis loops by modification of process variables

Loop	Case	Process Variables									Product
		P /bar	$T_{(F)}$ /K	T_{in} /K	$\dot{m}_{(3)}$ /kg h^{-1}	$Y_{H_2,(3)}$ /mol%	$Y_{N_2,(3)}$ /mol%	$Y_{NH_3,(3)}$ /mol%	$Y_{Ar,(3)}$ /mol%	$\dot{m}_{(4)}$ /kg h^{-1}	$\dot{m}_{(7)}$ /kg h^{-1}
I	MIN	220	550	623	372	15.88	78.54	3.56	2.02	–	12
	MAX	220	550	699	777	79.25	16.33	3.10	1.32	–	196
II	MIN	220	550	623	457	15.23	74.90	7.96	1.91	444	12
	MAX	220	550	697	955	71.48	13.92	12.77	1.83	783	173

The above shared load ranges of ammonia production are determined after maintaining the balance between heat production (reactor) and heat removal (heat exchanger) in the reactor system. For this purpose, the optimiser performed the stability analysis and shared results for the best possible operational intersection points among heat production and heat removal curves, see figure 4.6. For both loops' reactor systems at normal and maximum ammonia production, it can be seen that heat production curves and heat removal lines intersect at three points, whereas for minimum ammonia production, they intersect at only one point. For the cases of minimum and maximum ammonia production, heat removal lines are identical for both synthesis loops, as the highest possible process feed temperature is selected by the optimiser, see table 4.5. Heat production curves of the reactor system for minimum and maximum ammonia production varied significantly: for minimum ammonia production the intersection point is at the lowest possible inlet temperature 623 K. While for maximum ammonia production the highest possible inlet temperature of 699 K is selected and this highest temperature is implemented due the limitation of catalyst (Fe_3O_4), as at outlet of bed 1, 803 K temperature is approached, see figure 4.7a. Replacing the iron-based catalyst with a higher temperature resistance catalyst in bed 1 can be an option for increasing the ammonia production further.

In figure 4.7, the temperature profiles in the catalyst beds of the reactor system of the synthesis loops are presented. For minimisation, it can be observed that temperature becomes constant near the end of bed 3, as reaction equilibrium is achieved in the reactor systems of the synthesis loops. The lowest and the highest possible operational temperatures are achieved at the inlet and outlet of catalyst bed 1 for the minimum and maximum ammonia production, respectively,

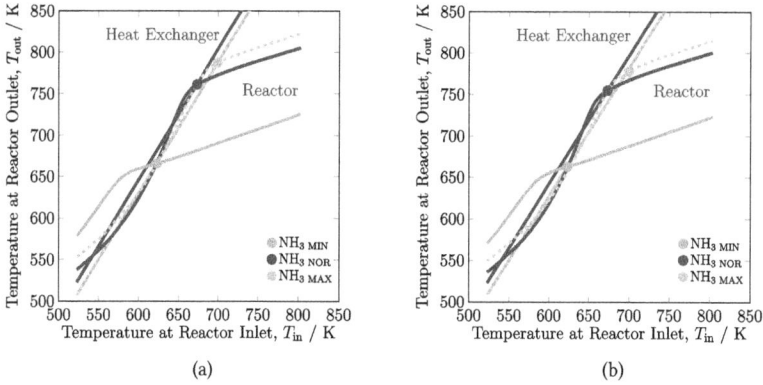

Figure 4.6.: Steady-state characteristics of the reactor (S-shaped curves) and the heat exchanger (straight lines) for normal (NOR), minimum (MIN) and maximum (MAX) NH_3 production: (a) the ammonia separator after the reactor system (loop I) and (b) the ammonia separator before the reactor system (loop II).

for the synthesis loop I, see figure 4.7a. Whereas, the lowest temperature, 623 K is achieved in catalyst bed 1 for synthesis loop II, see figure 4.7b. The possible reasons for not achieving the highest temperature 803 K in catalyst bed 1 is the higher ammonia content in the process feed, lower H_2 concentration in process feed and lower inlet temperature compared to loop I, which resulted in lower reaction rate.

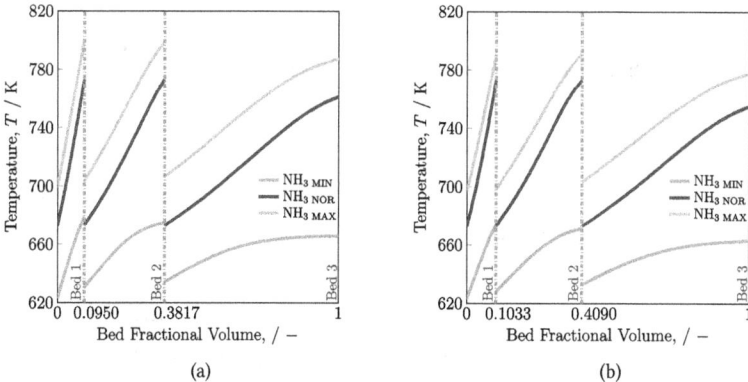

Figure 4.7.: Temperature profiles in the catalyst beds of the reactor systems for normal (NOR), minimum (MIN) and maximum (MAX) NH_3 production: (a) the ammonia separator after the reactor system (loop I) and (b) the ammonia separator before the reactor system (loop II).

In figure 4.8, the reactants conversion profiles along the catalyst beds of the reactor system of the synthesis loops I and II are presented. For minimum ammonia production, in reactor system of loop II lower H_2 conversion can be seen with respect to reactor system of loop I. From this it can be concluded that synthesis loop II requires slightly more H_2 intake than loop I, which is also evident from figure 4.9.

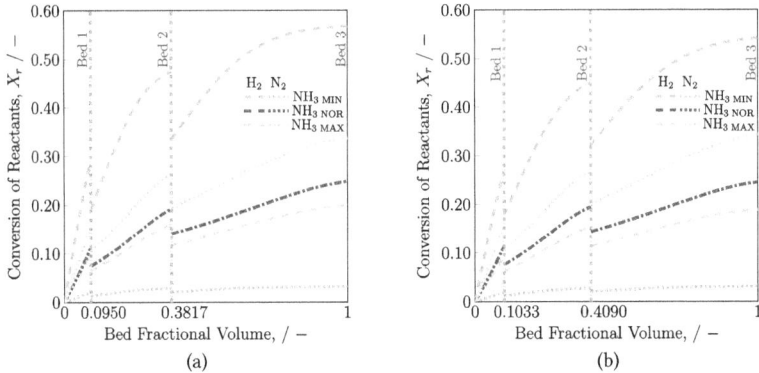

Figure 4.8.: Reactants conversion profiles in the catalyst beds of the reactor systems for normal (NOR), minimum (MIN) and maximum (MAX) NH_3 production: (a) the ammonia separator after the reactor system (loop I) and (b) the ammonia separator before the reactor system (loop II).

Finally, table 4.5 reveals that synthesis loops I and II are able to operate at same minimum load of 12 kg/h. This is in line with figure 4.9, which shows that the flexibilities for H_2 intake and recycle load also differ only slightly from each other. However, purge ratio reaches 10 %, *i.e.* a 400 % increase from the normal operational value for loop I, but only 180 % increase for loop II. For maximum ammonia production, loop II prefers not to operate at maximum purge ratio (10 %), mainly to avoid ammonia losses, as the purge stream is located immediately after the reactor system (see figure 4.1b), and at this location of the synthesis loop II, ammonia content is maximum (see figures B.13b, B.14b and B.15b (appendix B)). The reallocation of the purge stream, to immediately after the separator, could be beneficial for minimising ammonia losses. However at this location of the purge stream, reactants loss will increase as the reactants content is maximum, and also it is quite possible that loop II will be not able to achieve minimum load of 12 kg/h, as ammonia loss will decrease. This can be evident by comparing the percentage yields of the two loops (see figure 4.10). For example, for minimum NH_3 production, the percentage yield of the loop II is 10 % lower than of loop I. In general, from the percentage yield analysis of the two loops, it can be revealed that along with the mass flow rate of the purge stream, the allocation of the purge stream is also important for maximising the efficiency of the synthesis loops (see figures B.13 to B.15 (appendix B)). Furthermore, from the flexibility analysis for maximum ammonia production, a notable differences between the net product flexibilities of two loops are observed, where synthesis loop I provides *ca.* 18 % more ammonia production flexibility. However, this ammonia production flexibility comes at the cost of more H_2 intake. As a consequence, the purge stream flow rate (pure ratio) increases and thus the losses as a result lower the percentage yield and the recycle to fresh feed ratio decreases by 50 %.

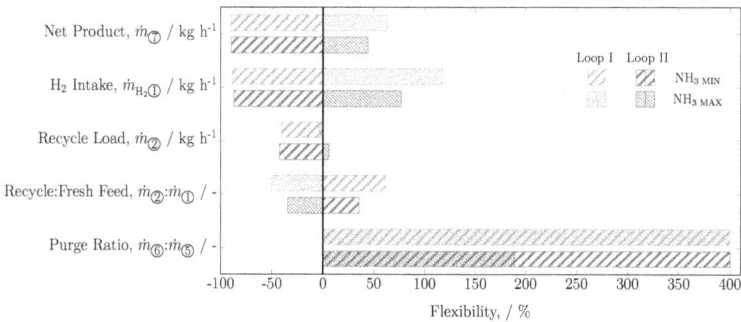

Figure 4.9.: Net product, H_2 intake, recycle load, recycle-to-feed ratio and purge ratio flexibilities in the ammonia synthesis loops I and II.

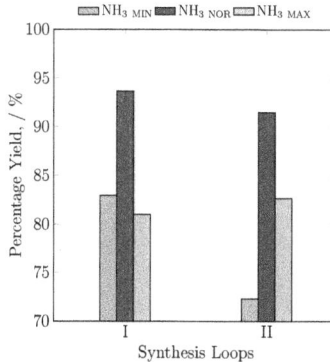

Figure 4.10.: Percent yield[1] of the ammonia synthesis loops I and II.

---

[1]The percentage yield is calculated by using equation A.36 (appendix A). Generally, in commercial ammonia plants an overall conversion up to 97 % is achievable. [24]

4.4. Conclusions

From the resulting load ranges of the synthesis loops I and II, it can be concluded that synthesis loop I is able to operate at a larger load range. Also, the synthesis loop I is more economical than the synthesis loop II in terms of CAPEX and OPEX. As from table 4.3, it can be seen that reactor volume and handling feed flow rate are lower, and hence compressing cost is reduced. In addition, for low flow rates, it will require an ammonia separator of lower volume, lower area for heat exchange and therefore less dependency on utilities for ammonia cooling and separation. In addition, by comparing flexibility results (loop I) of this chapter with chapters 2 and 3 (see table 2.3 and figure3.9), it can also be concluded that by modifying multiple process variables, overall NH_3 production load range enhances by almost two times. In single process variable modification, for minimising NH_3 production, H_2-to-N_2 ratios were beneficial and for maximising NH_3 production, Ar gas removal from the synthesis loop was beneficial. However, with multi-variable optimisation, minimum load further reduced by 15 % and maximum load enhanced by 40 %.

5. Load range enhancement of Haber-Bosch process designs

Capacity flexibility is known as the ability to adapt production output with regard to a change in market demands, and is further divided into two categories: volume flexibility and expansion flexibility.[65] For the chemical-based plants: volume flexibility relates to the operational flexibility and expansion flexibility relates with the installation of additional production lines.[66] With regard to power-to-ammonia applications, flexibility of both types is advantageous, as ammonia production is dependent on renewable energy, and renewable energy is seasonal, intermittent and decentralised when harvested. On the other hand, for the conventional Haber-Bosch process-based ammonia production, expansion flexibilities are well known[24], however operational flexibilities (load ranges) were not of interest due to the abundant availability of fossil fuel. Therefore, conventional synthesis loops are designed for operation at stable conditions and the load range of the synthesis loop is hard to quantify.[25] Hence, the focus of this work is to determine the operating flexibility of Haber-Bosch process designs. Generally, operational flexibility relates to operating range of the parameters, under which product quality does not change, e.g. product purity.[67,68]

In this work, so far operational and production (capacity) flexibilities of Haber-Bosch process designs for power-to-ammonia have been considered. The operating envelope of the individual process variables (chapter 2), the load range of the ammonia production by varying individual process varables (chapter 3) and multiple-variables (chapter 4) have been examined. Additionally, operational and production flexibilities of the power-to-ammonia process while knowing the reactants intake flexibility will be also of interest, as production of the reactants H_2 and N_2 is dependent on the renewable energy supply. Specially, H_2 production is strongly affected by changes in renewable energy, as more than 90 % of the energy is consumed during production of H_2 via electrolysis of water.[24] Therefore, during renewable energy shortages or outage periods, the minimum H_2 intake becomes of great importance.

Thus, in this chapter, along with ammonia production flexible analysis (minimisation and maximisation) for various Haber-Bosch process designs, minimum H_2 intake will also be considered. In chapter 3, five reactor system configurations (see figure 3.1) which differ in terms of inter-stage cooling methods were systematically analysed and compared regarding design and off-design performances. For design-based performance analyses, process feed ③ composition for all of the reactor systems was maintained by allowing variations in the purge ratio of the synthesis loop (see table 3.3). Due to this fact, the equilibrium and operational line for all design variants remained the same. It is pertinent to mention that to maintain the process feed ③ composition, net product flow rate was adjusted accordingly, without taking the separation model into account. The net product was considered to be $100\%\,NH_3$, thus for the design variants, along with an increase in ammonia production, an ammonia separator with higher efficiency was required (see table 3.4). Furthermore, for off-design analysis, ammonia production optimisation (minimisation and maximisation) was implemented by variation of one process variable at a time, *i.e.* argon composition Y_{Ar} or process feed flow rate $\dot{m}_{③}$ or H_2-to-N_2 ratio. In chapter 4, the synthesis loops I (separator after reactor system) and II (separator before reactor system), shown in figure 4.1, are compared for ammonia production flexibilities by multi-variable optimisation. The comparisons between the synthesis loops were only made for the direct cooling by quenching of reactor system 2Q. In addition, a NH_3 separator model was implemented along with the synthesis loop material balance. From the findings of chapter 4, it was concluded that the synthesis loop I is capable of handling more ammonia production load. By comparing the flexible ammonia production results of chapters 3 and 4 for reactor system 2Q, see figure 3.9 and 4.9, respectively, it can be concluded that with implementation of multi-variable optimisation, the ammonia production flexibilities enhances by more than 50% from single-variable optimisation. Therefore, in this chapter, the methodology of chapters 3 and 4 is combined for enhancing the load range of the Haber-Bosch reactor systems and the synthesis loop I shown in figures 3.1 and 4.1a, respectively. In this chapter, the main focus is to compare reactor systems flexibilities, with regard to NH_3 production load range (minimisation and maximisation) and H_2 intake (minimisation) by implementation of multi-variable optimisation.

5.1. Methodology

In this section, formulation of the optimisation problem for design and off-design performances of the five design variants shown in figure 3.1 with regard to synthesis loop I (figure 4.1a) are discussed. Also, the strategy for implementation in MATLAB is stated.

5.1.1. Problem formulation

For comparing the performances among design variants in the loop I load ranges, *i.e.* normal, minimum and maximum NH_3 production rates, and minimum H_2 intake are considered. The reactor system 2Q (direct cooling by quenching) is taken as a reference for other design variants. Models related to the reactor system, separation unit and overall ammonia synthesis loop are taken from previous chapters. For example, the model for the catalyst bed and the mixer in the reactor system is taken from chapter 2. The model for the heat exchanger in reactor system is taken from chapter 3 and the model of the ammonia separation unit is taken from chapter 4. The model for the overall synthesis loop, and the model related to the mixer and the splitter of the synthesis loop is taken from chapter 2.

Design performance

For the design performance comparison among five design variants, the volume of the catalyst beds and the fresh feed flow rate of the reference reactor system (2Q) for the synthesis loop I are taken from tables 4.3 and 4.4, and summarised in table 5.1. In contrast to chapter 4, production is optimised for fixed geometry to keep systems comparable. Furthermore, for the design performance comparison, initial and operational conditions, such as fresh feed composition, reactants conversion, process feed temperature and synthesis loop pressure are taken from table 2.1 and kept constant.

Table 5.1.: Fresh feed flow rate and catalyst bed volumes

Flow rate of fresh feed	Volume of catalyst beds			
$\dot{m}_{①}$ / kg h^{-1}	V_{b1} / m^3	V_{b2} / m^3	V_{b3} / m^3	$V_{b\,Total}$ / m^3
131.11	0.0069	0.0208	0.0448	0.0725

The optimisation problem (equation 5.1) of identifying the maximum net product rate $\dot{m}_{⑦,RS}$ for the reactor systems $RS \in \{HQ, QH, 2H\text{-}2, 2H\text{-}3\}$ is implemented using the equality and non-equality constraints. Equations 2.2 to 2.6 (chapter 2), 3.2a and 3.2b (chapter 3) are related to the reactor system, and equations A.32 to A.34, A.22 and A.24 (appendix A) are related to the synthesis loop. For the ammonia separation unit, equations 4.1 to 4.4 are taken from chapter 4, however subscript L must be replaced with RS.

$$\underset{z_{RS}}{\text{maximise}} \ \ \dot{m}_{\text{⑦},RS} \tag{5.1}$$

subject to

Equations 2.2 to 2.6 (reactor system).

Equations 3.2a and 3.2b (reactor system).

Equations 4.1 to 4.4 (separation unit).

Equations A.32 to A.34 (synthesis loop).

Equations A.22 and A.24 (synthesis loop).

The reactor systems $RS \in \{$HQ, QH, 2H-2, 2H-3$\}$ contain interstage mixer(s), heat exchanger(s) or a combination of both, see figure 3.1. Therefore, manipulatable variables change accordingly, which results in the following variables z_{RS} of respective reactor systems: $z_{HQ} \in \{T_{b1_{in}}, T_{b2_{in}}, \dot{m}_{b1_{in}}, \dot{m}_{q3}\}$ $z_{QH} \in \{T_{b1_{in}}, T_{b2_{in}}, \dot{m}_{b1_{in}}, \dot{m}_{q2_{in}}, Y_{c,\text{③}}\}$, $z_{2H-2} \in \{T_{b1_{in}}, T_{b2_{in}}, T_{b3_{in}}, \dot{m}_{b1_{in}}, Y_{c,\text{③}}\}$ and $z_{2H-3} \in \{T_{b1_{in}}, T_{b2_{in}}, T_{b3_{in}}, \dot{m}_{b1_{in}}, Y_{c,\text{③}}\}$, where $c \in \{N_2, H_2, NH_3, Ar\}$. Like previous chapters, for design performance, the H_2-to-N_2 molar ratio is maintained at $3 : 1$, $\sum_{c=1}^{4} Y_{c,\text{③},RS} = 1$ is assured and the inert gas concentration within the synthesis loop of the design variants is controlled by maintaining a purge ratio $p_{RS} = 0.02$.

For heat exchangers, knowing effectiveness $\varepsilon_{HE,RS}$ at design conditions allows one to determine the unknown exit stream temperatures during off-design performance analysis. The effectiveness of heat exchangers $HE \in \{1, 2, 3\}$ of the reactor systems $RS \in \{$HQ, QH, 2H-2, 2H-3$\}$ are calculated by either equation 3.1a or 3.1b for the optimum design temperatures. Equation 3.1a applies to $HE \in \{1, 2, 3\}$ of the reactor systems 2H-2 and 2H-3, and to HE 2 of the HQ. For HE 1 of the reactor system HQ and two heat exchangers of the reactor system QH, equation 3.1b is used.

The boundary limitations of the process variables for the design performance analysis of the reactor systems are given in table 5.2. The process feed composition boundaries are similar to table 4.1, however for mass flow rates, the upper boundary is doubled, as the process feed for reactor systems HQ, QH, 2H-2 and 2H-3 sub-divide in fewer than three streams. The inlet temperature boundary for the catalyst beds is implemented with regard to preferable operational range (equation 4.9a).

Table 5.2.: Boundary limits of the process variables and the design parameters for the design performance analysis of the reactor systems

Process variables	Lower bound	Upper bound
$T_{b_{in},RS}$ / K	673	773
$\dot{m}_{b1,RS}$ / kg h^{-1}	200	800
$\dot{m}_{q2,RS}$ / kg h^{-1}	100	400
$\dot{m}_{q3,RS}$ / kg h^{-1}	150	500
$Y_{H_2,③,RS}$ / mol %	0	100
$Y_{N_2,③,RS}$ / mol %	0	100
$Y_{NH_3,③,RS}$ / mol %	0	100
$Y_{Ar,③,RS}$ / mol %	0	15

Off-design performance

For the off-design performance analysis and comparison among design variants, minimum and maximum NH_3 production, and minimum H_2 intake limitations are considered. Operation of the reactor design variants is considered to be autothermal. Therefore during off-design performance analysis, management between heat generation (reactor) and heat removal (heat exchanger) is also necessary. The material and energy balance equations 2.2 to 2.6 of the reactor system for design and off-design performance remain the same. However, equations 3.2a and 3.2b related to the inequality constraints in catalyst bed are replaced with equation 3.3. In addition, for heat exchangers within reactor systems, equality constraint equations 3.1a and 3.1b are implemented to determine the temperature of the unknown exit stream. Also within heat exchangers, the inequality constraint equation 3.4 for minimum temperature difference is implemented to guarantee feasible operation. For off-design performance, the equations related to the material balance in the synthesis loop (equations A.32 to A.34, A.22 and A.24) and ammonia separation unit (equations 4.1 to 4.4) remain similar to design performance. The optimisation problems for minimum and maximum NH_3 production are formulated as follows:

$$\underset{z_{RS}}{\text{minimise}} \mid \text{maximise } \dot{m}_{⑦,RS} \tag{5.2}$$

subject to

Equations 2.2 to 2.6 (reactor system).

Equation(s) 3.1a and/or 3.1b (reactor system).

Equations 3.3 and 3.4 (reactor system).

Equations 4.1 to 4.4 (separation unit).

Equations A.32 to A.34 (synthesis loop).

Equations A.22 and A.24 (synthesis loop).

and for minimum H_2 intake as:

$$\underset{z_{RS}}{\text{minimise}} \ \dot{m}_{H_2,\textcircled{1},RS} \tag{5.3}$$

subject to

Equations 2.2 to 2.6 (reactor system).

Equation(s) 3.1a and/or 3.1b (reactor system).

Equations 3.3 and 3.4 (reactor system).

Equations 4.1 to 4.4 (separation unit).

Equations A.32 to A.34 (synthesis loop).

Equations A.22 and A.24 (synthesis loop).

where in equation 5.2 $\dot{m}_{\textcircled{7},RS}$ is the net product and in equation 5.3 $\dot{m}_{H_2,\textcircled{1},RS}$ is the H_2 intake of the synthesis loop for the design variants $RS \in \{2Q, HQ, QH, 2H\text{-}2, 2H\text{-}3\}$. For design variants, $z_{RS} \in \{P_{RS}, T_{\textcircled{3},RS}, T_{in,RS}, \dot{m}_{\textcircled{3},RS}, Y_{c,\textcircled{3},RS}\}$ and $z_{2H\text{-}3}$ contains $T_{cold_{out}, HE3}$ in addition. $\dot{m}_{\textcircled{3},RS}$ is the mass flow rate of the process feed, and it may subdivide into several streams (see figure 3.1), however the change in flow rate is proportional for all streams. Furthermore, for minimisation $Y_{H_2,\textcircled{3},RS} < Y_{N_2,\textcircled{3},RS}$ and for maximisation $Y_{H_2,\textcircled{3},RS} > Y_{N_2,\textcircled{3},RS}$ need to be assured in the process feed, and also $\sum_{c=1}^{4} Y_{c,\textcircled{3},RS} = 1$. In comparison to design performance, the purge ratio p_{RS} is set free with boundary conditions implementation, i.e. $0 < p_{RS} \leq 0.10$.

It is pertinent to mention that in this chapter for the reference reactor system 2Q, only minimisation of H_2 is done and the results of ammonia load range optimisation are taken from chapter 4. For off-design performances, boundary limits of the process variables for the reactor systems HQ, QH, 2H-2 and 2H-3 are similar to the reactor system 2Q (table 4.2). However, for off-design optimisation of reactor system 2H-3, boundaries for $T_{cold_{out}, HE3, 2H\text{-}3}$ are also required. Therefore along with it, all boundaries are stated again in table 5.3. For $T_{cold_{out}, HE3, 2H\text{-}3}$, the maximum possible temperature span in considered, i.e. lower bound of $T_{\textcircled{3},RS}$ and upper bound of $T_{in,RS}$.

Table 5.3.: Boundary limits of the process variables for the off-design performance analysis of the reactor systems

Process variables	Lower bound	Upper bound
P_{RS} / bar	150	220
$T_{③,RS}$ / K	450	550
$T_{in,RS}$ / K	623	773
$\dot{m}_{③,RS}$ / kg h^{-1}	0.55 $\dot{m}_{③,NOR,RS}$	1.15 $\dot{m}_{③,NOR,RS}$
$Y_{H_2,③,RS}$ / mol %	0	100
$Y_{N_2,③,RS}$ / mol %	0	100
$Y_{NH_3,③,RS}$ / mol %	0	100
$Y_{Ar,③,RS}$ / mol %	0	15
$T_{cold_{out}, HE3, 2H-3}$ / mol %	450	773

5.1.2. Problem implementation

Like chapters 3 and 4, implementation of optimisation problems are carried out in MATLAB by using the fmincon solver, and multiple initialisations with several initial values were carried out. The optimisation scheme for design and off-design performance is adapted similar to the loop I (see figures 4.3a and 4.4a), however process variables vary, and also the objective function of design performance problem.

5.2. Results and discussions

The results obtained for the optimised design and off-design performances of the reactor systems are presented and discussed in this section. First, results regarding the design performance of the reactor systems for maximum net production are presented and discussed. Afterwards, optimisation results regarding off-design scenarios, i.e. minimum and maximum net production (NH_3) and minimum H_2 intake are presented and discussed. Finally, a flexibility analysis is made for the reactor systems.

5.2.1. Design performance

Here, the results obtained for maximum net production of the reactor systems $RS \in \{HQ, QH, 2H-2, 2H-3\}$ by varying the process variables $z_{HQ} \in \{T_{b1_{in}}, T_{b2_{in}}, \dot{m}_{b1}, \dot{m}_{q3}, Y_{c,③}\}$, $z_{QH} \in \{T_{b1_{in}}, T_{b2_{in}}, \dot{m}_{b1}, \dot{m}_{q2}, Y_{c,③}\}$, $z_{2H-2} \in \{T_{b1_{in}}, T_{b2_{in}}, T_{b3_{in}}, \dot{m}_{b1}, Y_{c,③}\}$ and $z_{2H-3} \in \{T_{b1_{in}}, T_{b2_{in}}, T_{b3_{in}}, \dot{m}_{b1}\}$ together are presented, then for design performance of the reactor systems, temperature-reactants conversion trajectories and effectiveness of the heat exchangers in the reactor systems are presented.

From table 5.4 it can be seen that the reactor systems 2H-2 and 2H-3 result in the highest net product and are also optimised for the same process variable values. This finding is quite similar to the chapter 3 (table 3.2). Both reactor systems handle the lowest process feed flow rate, but resulted in the highest net product production. The primary reason is that the whole process feed passes through all catalyst beds, and therefore has more catalyst volume for the reaction and also more space time. In addition, from the handling of less flow rate, visible in the sum of mass flows $\dot{m}_{\textcircled{3}}$, it can also be concluded that the recycle load is minimal for 2H-2 and 2H-3 in comparison to HQ and QH, as fresh feed composition and flow rate are fixed for all reactor systems (see tables 2.1 and 5.1). Furthermore, it can be concluded that the reactor systems 2H-2 and 2H-3 require less reactants intake, as well as less NH_3 recycling, but prefer to maintain a slightly higher inert gas concentration within the synthesis loop.

Table 5.4.: Optimal process variables (highlighted in grey colour) for the reactor systems and resulting maximum net product rate (enclosed in rectangle)

Process variables	Reactor Systems			
	HQ	QH	2H-2	2H-3
$T_{b1_{in}, NOR}$ / K	689.86	683.84	698.07	698.07
$T_{b2_{in}, NOR}$ / K	680.54	680.75	686.03	686.03
$T_{b3_{in}, NOR}$ / K	674.05	675.96	676.64	676.64
$\dot{m}_{b1, NOR}$ / kg h^{-1}	492.67	431.20	621.55	621.55
$\dot{m}_{q2, NOR}$ / kg h^{-1}	-	202.18	-	-
$\dot{m}_{q3, NOR}$ / kg h^{-1}	169.08	-	-	-
$\dot{m}_{\textcircled{3}, NOR}$ / kg h^{-1}	661.75	633.38	621.55	621.55
$Y_{H_2, \textcircled{3}, NOR}$ / mol %	67.70	67.50	67.41	67.41
$Y_{N_2, \textcircled{3}, NOR}$ / mol %	22.57	22.50	22.47	22.47
$Y_{NH_3, \textcircled{3}, NOR}$ / mol %	3.65	3.58	3.55	3.55
$Y_{Ar, \textcircled{3}, NOR}$ / mol %	6.08	6.41	6.56	6.56
$\dot{m}_{\textcircled{7}, NOR}$ / kg h^{-1}	120.28	120.86	121.10	121.10

The temperature-reactants conversion $\overline{TX}$ trajectories of the reactor systems are shown in figure 5.1. It can be seen that all the reactor systems operated for maximum conversion i.e. 90 % of the equilibrium conversion. Among all reactor systems, 2H-2 and 2H-3 result in highest reactants conversions which is due to lower feed flow rate $\dot{m}_{\textcircled{3}, NOR}$ and reactants concentration (see table 5.4), and due to the highest space time, as the process feed passes through all catalyst beds (see figure 3.1). In contrast to chapter 3 (see figure 3.4), here the equilibrium (EQ) and operation (OP) line of the reactor systems HQ, QH and 2H (2H-2 and 2H-3) vary, as process feed composition for all the reactor system varies from each other, except 2H-2 and 2H-3, see table 5.4.

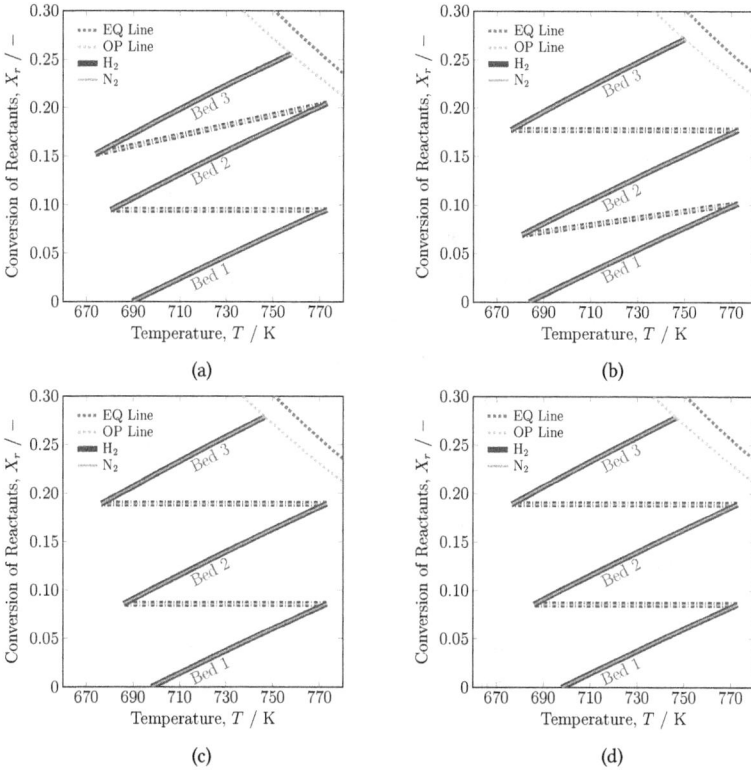

Figure 5.1.: Temperature - reactants conversion $\overline{TX}$ trajectories of the reactor systems: (a) HQ, (b) QH, (c) 2H-2 and (d) 2H-3 along with equilibrium (EQ) and operational (OP) lines.

The effectivenesses of the heat exchangers are calculated for the design performance of the reactor systems by means of equation 3.1a or 3.1b, and the resulting values are presented in table 5.5. During off-design performance analysis, these effectivenesses of the heat exchangers will be used for calculating the temperature of a required stream.

Table 5.5.: Resulting effectiveness of heat exchangers for optimised reactor systems

RS	$\varepsilon_{HE\,1}$ / -	$\varepsilon_{HE\,2}$ / -	$\varepsilon_{HE\,3}$ / -
HQ	0.5462	0.3698	-
QH	0.3456	-	0.5032
2H-2	0.1867	0.3479	0.5682
2H-3	0.1867	0.5243	0.3854

In comparison to chapter 3, the optimum design performance results obtained for the inlet temperatures of the catalyst beds and the flow rates of the quench streams (see table 5.4) for the reactor systems are quite similar, see table 3.2. However, in chapter 3, process feed flow rate and composition was considered fixed for all the reactor systems, as the model of the ammonia separator was not taken into account, thus all reactor systems were able to operate for the same operational line, see figure 3.4. Whereas, in this chapter, beside the reactor system model, the ammonia separator model is also considered. Therefore, the process feed composition and flow rate are also set free (see table 5.4), and thus all the reactor systems have different operational lines, see figure 5.1. In general, all the reactor systems in this chapter resulted in higher reactants conversion than the chapter 3, this is due to lower process feed flow rate and low ammonia content. However, due to higher process feed flow rate and volume of catalyst beds, reactor systems of chapter 3 resulted in higher net product rate, see table 3.4. In fact, a more detailed comparison between the results of the reactor systems of both chapters is omitted here, as the basis of the synthesis loops are also not same.

5.2.2. Off-design performance

In this section, the results obtained for the flexibility analysis of the reactor systems by minimum and maximum net product and minimum H_2 intake are presented. The optimum values of the process variables and the temperature profiles for the reactor systems are presented. These profiles are presented for the intersection points between the heat production and removal curves for normal, minimum and maximum net product production, and minimum H_2 intake. Furthermore, the reactor systems flexibility analysis and comparison is made.

It can be seen in table 5.6 that all the reactor systems are similarly capable to operate for large load ranges of ammonia production, that is between 12 kg/h and 203 kg/h. Among all these, the process variables $Y_{H_2, \text{③}}$, $Y_{N_2, \text{③}}$, $Y_{Ar, \text{③}}$ and process feed flow rate $\dot{m}_{\text{③}}$ vary to a larger extent, which also resembles previous results of single process variable variations, see table 2.3 (chapter 2). For minimising and maximising net production, all the reactor systems prefer to operate at minimum and maximum process feed loads, respectively. For operation pressure it can be seen that for minimum and maximum net product production, the reactor systems HQ and QH selected the highest pressure, whereas for minimum net product, reactor systems 2H-2 and 2H-3 preferred to operate at pressures lower than the normal scenario (200 bar). Operation of the reactor systems at lower pressure for minimum load would be more economical, as compressors within the synthesis loop would have lower energy consumption. Furthermore, it can be seen that for process feed temperature, a higher boundary limit is achieved for minimising and maximising net product production cases of the reactor systems. In addition, it can be seen that for the minimum net product production case, the reactor systems HQ, 2H-2 and 2H-3 preferred to operate at the highest possible concentration of inert gas (15 %) within the synthesis loop, whereas reactor system QH did not prefer the highest inert gas concentration, but prefers to operate at the lowest inlet temperature. For minimum NH_3 production, higher inert content in the process feed (synthesis loop) seems more beneficial, as by maintaining higher Ar concentration in synthesis loop, purging will be carried out at a lower rate (see figure 5.5), which results in lower reactants emission. In general, through higher inert content in synthesis loop, the load of recycle stream must also increase (see table 2.3).[6] However, from figure 5.5 it is observed that load of recycle stream declined. Actually, along with Ar concentration in process feed, mass flow rate has also decreased and therefore resulted in less load on the recycle stream (see table 2.3). Whereas, for reactor systems 2Q pressure has also increased, which is also supportive of lowering recycle load (see table 2.3). In general, low mass flow rate and higher pressure enhances reactants conversion, see table B.2. For multivariable modification, figuring out individual variable effect and impact is not easy, as other variables also change and the effect may neutralise.

From the results of H_2 intake minimisation given in table 5.7, it is observed that the process variables pressure, feed temperature, reactor inlet temperature and process feed flow rate resulted in the same optima as of minimum NH_3 production, see tables 5.6 and 4.5. However, variation in feed composition can be noticed, especially for the reactor systems 2Q and HQ, as

Table 5.6.: Results of the reactor systems optimisation for NH_3 production (equation 5.2) by modification of process variables

RS	Off-Design	Process Variables									Product
		P /bar	$T_{③}$ /K	T_{in} /K	$\dot{m}_{③}$ /kg h⁻¹	$Y_{H_2,③}$ /mol%	$Y_{N_2,③}$ /mol%	$Y_{NH_3,③}$ /mol%	$Y_{Ar,③}$ /mol%	$T_{cold_{out}, HE3}$ /K	$\dot{m}_{⑦}$ /kg h⁻¹
HQ	MIN	220	550	629	364	15.72	65.35	3.92	15.00	–	12
	MAX	220	550	716	761	78.86	16.71	3.07	1.36	–	196
QH	MIN	220	550	623	348	15.12	79.29	3.54	2.05	–	12
	MAX	220	550	710	728	78.73	16.83	3.00	1.44	–	201
2H-2	MIN	188	550	637	342	19.90	60.56	4.54	15.00	–	14
	MAX	220	550	724	715	78.43	17.12	2.97	1.48	–	201
2H-3	MIN	157	550	638	342	22.15	57.33	5.52	15.00	608	12
	MAX	220	550	734	715	81.33	14.29	3.06	1.32	655	203

minimum H_2 intake for both preferred maximum Ar content. This is advantageous, as during purging, lower amount of reactants will be wasted or sent to the recovery section. Due to higher content of Ar, both reactors also preferred slightly higher H_2 content in the process feed, which helped in maintaining the same temperature output of the reactor system, see figure 5.2b.

Table 5.7.: Results of the reactor systems optimisation for H_2 intake (equation 5.3) by modification of process variables

RS	Off-Design	Process Variables									H_2 Intake
		P /bar	$T_{③}$ /K	T_{in} /K	$\dot{m}_{③}$ /kg h⁻¹	$Y_{H_2,③}$ /mol%	$Y_{N_2,③}$ /mol%	$Y_{NH_3,③}$ /mol%	$Y_{Ar,③}$ /mol%	$T_{cold_{out}, HE3}$ /K	$\dot{m}_{H_2,①}$ /kg h⁻¹
2Q	MIN	220	550	623	372	16.34	64.74	3.92	15.00	–	2.70
HQ	MIN	220	550	629	364	15.72	65.35	3.92	15.00	–	2.49
QH	MIN	220	550	623	348	15.78	65.32	3.90	15.00	–	3.25
2H-2	MIN	188	550	637	342	19.90	60.56	4.54	15.00	–	3.25
2H-3	MIN	158	550	637	342	21.92	57.60	5.47	15.00	608	2.20

The figures 5.2 and 5.3 show profiles of temperature and reactants conversion, respectively, along the catalyst beds of the reactor systems for the desired intersection points between the heat production and heat removal curves. Generally, it can be observed that the rise in temperature and reactants conversion occurs at a much higher rate in bed 1 than in beds 2 and 3. This is due to the low ammonia content presence; in addition bed 3 is handling more feed.

From figure 5.2, for minimum NH_3 production and H_2 intake of the reactor systems 2H-2 and 2H-3, minimum temperature difference (pinch point) is approaches in the heat exchanger 1. From this, it can be stated that for minimisation of NH_3 and H_2 intake, both reactor systems utilise heat of reaction sufficiently and effectively. In reactor systems QH, 2H-2 and 2H-3, for maximum NH_3 production at the exit of bed 1, the maximum temperature of 803 K is approached, which indicates that these reactor systems are performing at their highest possible level and the maximum possible reactants conversion has been approached.

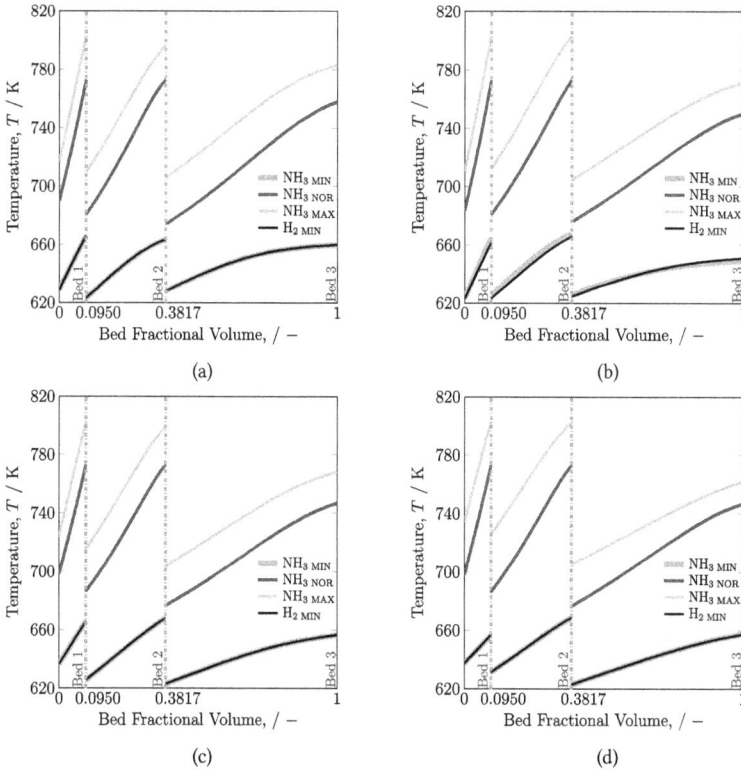

(a)

(b)

(c)

(d)

Figure 5.2.: Temperature profiles for normal (NOR), minimum (MIN) and maximum (MAX) NH_3 production, and minimum (MIN) H_2 intake of the reactor systems: (a) HQ, (b) QH, (c) 2H-2 and (d) 2H-3.

From the figure 5.3, it can be observed that the reactants conversion is inversely proportional to the reactants concentration in the process feed, see table 5.6. Additionally, the effect of the inter-stage quenching and the heat exchanger can also be distinctly observed. For example, continuity of reactants conversion between two beds indicates the presence of an inter-stage heat exchanger. For minimum NH_3 production, it can be observed that the H_2 conversion at the end of bed 3 of the reactor system QH is higher compared to the reactor system HQ.

This is due to less H_2 content in the process feed, and vice versa for minimum H_2 intake. Among all reactor systems, it can be observed that for minimum NH_3 production and H_2 intake, reactor system 2H-3 resulted in the lowest H_2 conversion. The reasons behind this lowest H_2 conversion are the lowest operational pressure and the highest H_2 concentration in the process feed, see table 5.6.

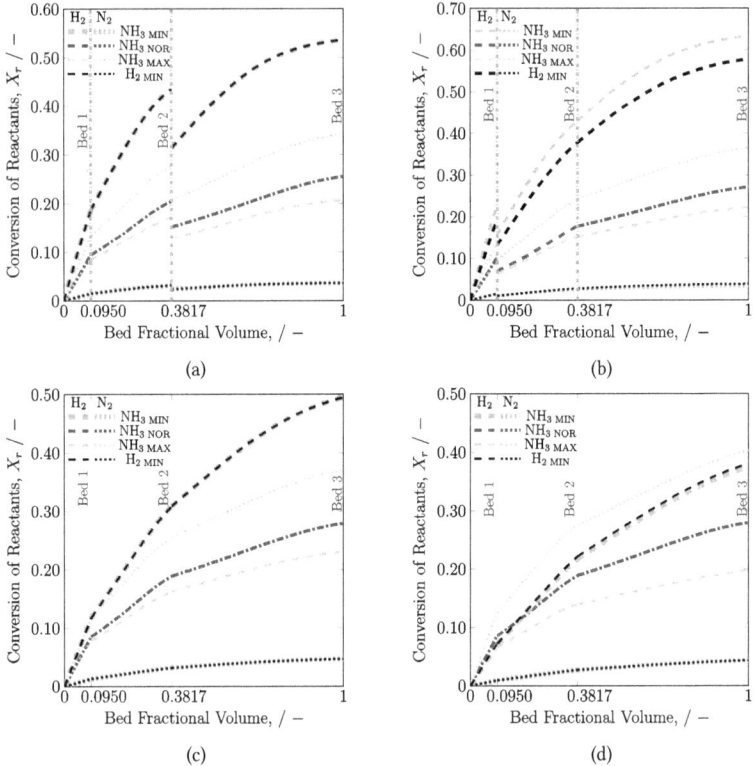

Figure 5.3.: Temperature profiles for normal (NOR), minimum (MIN) and maximum (MAX) NH_3 production, and minimum (MIN) H_2 intake of the reactor systems: (a) HQ, (b) QH, (c) 2H-2 and (d) 2H-3.

From the obtained results of optimisation, it seems that all the reactor systems are equally capable to operate within a large NH_3 production range and minimum H_2 intake. However, in parallel it is also important to know how efficiently these reactor systems are operating within the synthesis loop. For this, percentage yield analysis and flexible analysis of the synthesis loops are performed. From the values of percentage yields displayed in figure 5.4 it can be seen that for minimisation of H_2 intake, all the reactor systems resulted in maximum yield (*i.e.* around 99 %). In addition, reactor systems HQ, 2H-2 and 2H-3 also resulted in a quite similar percentage yield for minimum NH_3 production. On the other hand, for the reactor systems 2Q and QH, the difference of around 15 % can be observed between the percentage yields of the minimum NH_3 production and H_2 intake. For these two reactor systems, between minimum NH_3 production and H_2 intake, the only prominent difference is Ar concentration. So, for the minimisation, higher Ar content is beneficial. Furthermore, for maximum NH_3 production, the percentage yield of all the reactor systems decreases from the design performance. From the difference between the percentage yields of the reactor systems, it seems that NH_3 and reactants are also been lost, *e.g.* in purge stream. Therefore, in addition to percentage yield, the change in stream loads also needed to be evaluated.

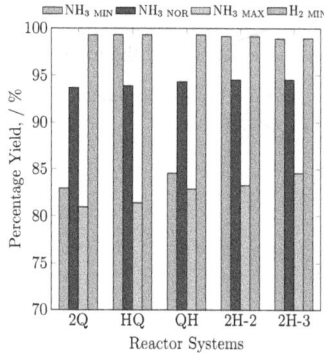

Figure 5.4.: Percent yield of the reactor systems 2Q, HQ, QH, 2H-2 and 2H-3 in synthesis loop.

In terms of flexibility, defined as fractional change from normal values, the reactor systems performances along with the synthesis loop can be seen in figure 5.5. All the reactor systems seem equally suitable for flexible ammonia production, *i.e.* 153 %, 154 %, 156 %, 155 % and 158 % for 2Q, HQ, QH, 2H-2 and 2H-3, respectively. In comparison to single-variable optimisation, multi-variable optimisation seems more beneficial for power-to-ammonia application, as it enhances load range of ammonia production more than two times, see figure 3.9. In addition, it can also be observed that for single process variable variations, reactor systems show strong differences in the permitted load variation, whereas for multi-variable, the load range of all reactor systems is quite similar. A change in ammonia production is directly proportional to a change in the H_2 intake and inversely proportional to a change in the recycle to fresh feed ratio. This is mainly due to the concentration of Ar and NH_3 (see tables 5.6 and 5.7). For example, higher content of Ar and NH_3 lowers reaction rate and, as a consequence, lower temperature within the catalyst beds, see figure 5.2. The second reason is the difference between the molecular weight of H_2 and N_2. For example, for NH_3 minimisation and maximisation in the reactor system QH, Ar and NH_3 content do not vary to a large extent (see table 5.6), but only H_2 and N_2 varied (see table 2.3 and figure 3.9c). In addition, it can be observed that for off-design operation, the recycle stream load will decrease. For maximum ammonia production, all the reactor systems preferred to have the maximum purge ratio, i.e. 10 % of recycle load. However, for minimum ammonia production, reactor systems HQ, 2H-2 and 2H-3 decrease their purge ratio by ca. 80 % from the normal value, i.e. 2 % of recycle load. For minimum hydrogen intake, all rector systems preferred to operate at a minimum purge ratio. In general, a lower purge ratio is more beneficial, as less product and reactants are wasted while regulating inert gas in the synthesis loop, which is also evident from the higher percentage yield, see figure 5.4. Therefore the decline in the purge ratio is more beneficial than the increase in purge ratio.

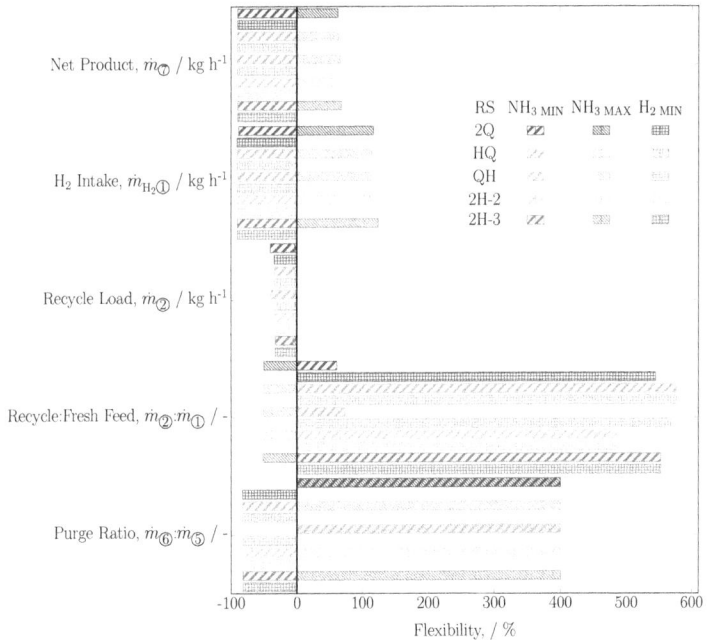

Figure 5.5.: Net product, H_2 intake, recycle load, recycle-to-feed ratio and purge ratio flexibilities in the ammonia synthesis loops of the reactor systems 2Q, HQ, QH, 2H-2 and 2H-3.

5.3. Conclusions

From the results of this chapter it can be concluded that by applying multi-variable modification, all the reactor systems 2Q, HQ, QH, 2H-2 and 2H-3 are capable of operation for large ranges of NH_3 production load. Also, when the objective function is switched from minimisation of NH_3 production to minimisation of H_2 intake, all the reactor systems resulted in higher yield *i.e.* above 99 %. Higher percentage yield, and lower load operation, *i.e.* up to 10 % of nominal load makes all the reactor systems equally suitable for renewable energy outage period application. At the same time, when renewable energy is excessively available, ammonia production can be enhanced nearly by 60 % with an excessive supply of H_2.

6. Summary, conclusions and outlook

The power-to-ammonia process allows for the production of ammonia, the second most produced inorganic chemical in the world, from water, air and renewable energy. With regard to the ammonia synthesis, the power-to-ammonia approach requires flexible operation and thus becomes applicable for buffering the intermittent renewable energy supply. The Haber-Bosch ammonia synthesis process has been intensively developed over a period of a century and has been optimised for stable operation only, due to the maximise economic conversion from fossil fuels. To achieve the requirement of flexible ammonia production and hydrogen intake, various Haber-Bosch ammonia synthesis loop configurations and reactor system designs have been introduced and analysed systematically by model simulation and optimisation. Special considerations are given to autothermal operation of synthesis reactor systems with a three-bed reactor system configuration with several inter-stage cooling methods.

6.1. Summary

Firstly, the operating envelope of the three-bed autothermal direct cooling by quenching (2Q) ammonia synthesis reactor system is investigated for various process variables, namely operational pressure, process feed temperature, process feed flow rate and process feed composition (*e.g.* Ar, NH_3 and H_2-to-N_2 ratio). Among these six process variables, Ar concentration, H_2-to-N_2 ratio and process feed flow rate provide the widest operating range. With regard to H_2 intake (fresh feed) and NH_3 production, H_2-to-N_2 ratio and Ar concentration in synthesis loop provide more flexibility. In particular, the operation of a synthesis reactor system at low H_2 intake could be quite useful during renewable energy outage periods for avoiding possible shut down.

Secondly, the three-bed autothermal ammonia synthesis reactor system is modified with regard to inter-stage cooling methods, such as: direct cooling by quenching (2Q), combinations of indirect and direct cooling (HQ and QH) and of indirect cooling (2H) with variations. Afterwards, these five design variants are compared for design and off-design performances. Minimum and maximum ammonia production scenarios are considered by varying one process variable at a time, *i.e.* inert gas percentage, process feed flow rate or H_2-to-N_2 ratio. For design performance analysis at fixed operational conditions, indirect cooling reactor systems (2H-2 and 2H-3) result in the highest NH_3 production. Whereas, for off-design performances, all five reactor systems seem quite viable. For example, with H_2-to-N_2 ratio variations, all were able to operate for significantly lower H_2 intake, *i.e.* two-thirds of design requirements.

Thirdly, two ammonia synthesis loop configurations are compared with considerations of the three-bed autothermal direct cooling by quenching (2Q) reactor system and the ammonia separation unit for design and off-design performance analyses. The synthesis loops differ in terms of ammonia separation unit allocation, *i.e.* after (loop I) and before (loop II) the synthesis reactor system. For the design performance analysis, constant fresh feed composition, operational pressure, process feed temperature and NH_3 production capacity are maintained. For off-design performance analysis, multi-variable optimisation is applied for minimum and maximum ammonia production. For design and off-design performances, loop I seems more significant, as it requires less reactor volume and resulted in higher percentage yield at design conditions, and for off-design, loop I is able to operate at a larger load range.

Fourthly, five reactor systems 2Q, HQ, QH, 2H-2 and 2H-3, along with the ammonia separation unit model, are implemented in the synthesis loop I for design and off-design performances. During design performance analysis, for the same reactor volume and process conditions of synthesis loop, reactor systems 2H-2 and 2H-3 produce the maximum amount of ammonia with minimum recycle load. For off-design analysis with implementation of multi-variable optimisation, all five reactor systems provided more than 150 % of flexibility in ammonia production and hydrogen intake. For minimum hydrogen intake, all the reactor systems performed quite similarly, and resulted in highest percentage yield, *i.e.* 99 %. However, for minimum ammonia production, the reactor systems performance vary; reactor systems HQ, 2H-2 and 2H-3 performed better in comparison to reactor systems 2Q and QH, as they preferred to operate with more than five times less purge ratio.

6.2. Conclusions

From the findings of this work, it can be concluded that all the Haber-Bosch process designs are capable of operation for a larger load range by changing several process variables simultaneously. For minimum ammonia production, all the synthesis loop configurations and the reactors system designs are able to operate ten times lower than the nominal production ca-pacity. Similarly, for minimum hydrogen intake, all the reactor systems are able to operate with ten times less hydrogen intake while having maximum percentage yield. Among synthe-sis loop configurations, the ammonia separation unit after the reactor system (loop I), provided the highest flexibility in ammonia production and hydrogen intake, and resulted in higher de-sign percentage yield. Low load operational ability of the Haber-Bosch process designs make them viable for the power-to-ammonia applications. For example, during renewable energy outage periods, the intermediate shut down of the plant can be avoided. On the other hand, operational ability of the Haber-Bosch process designs at higher load also make their usage li-able during renewable energy surplus periods. Therefore, all five reactor systems in the loop I configuration are recommended for power-to-ammonia applications.

6.3. Outlook

In this work, the implementation of material balances in general and energy balances, thermo-dynamic and kinetic models were considered. Detailed design analysis for the reactor systems and the synthesis loops have been not considered; therefore limitations which are forced by heat losses and fluid dynamics are neglected. For smaller scale plants and very low mass feed flow rates, heat losses might be noticeable and could influence the operating envelope. With consideration of design and construction specifications, along with site selection and envi-ronmental conditions, heat losses and fluid dynamics can be within the scope of future work. Furthermore, with consideration of design and operation of the heat exchangers and the cool-ers trail, along with the design of compressors, the economic aspect can also be included and the impact of the work can be enhanced.

Bibliography

[1] ISPT, Power to ammonia: Feasibility study for the value chains and business cases to produce CO_2-free ammonia suitable for various market applications, Technical report, Institute for Sustainable Process Technology, Amersfoort, The Netherlands (2017).

[2] G. L. Soloveichik, Ammonia for energy storage and delivery, in: The 13th Annual NH3 Fuel Conference, Los Angeles, CA, USA, Sep. 18-21, 2016.

[3] E. Morgan, J. Manwell, J. McGowan, Wind-powered ammonia fuel production for remote islands: A case study, Renew Energy 72 (2014) 51–61.

[4] J. Fuhrmann, M. Hülsebrock, U. Krewer, Energy storage based on electrochemical conversion of ammonia, in: D. Stolten, V. Scherer (Eds.), Transition to renewable energy systems, Wiley-VCH Verlag GmbH & Co. KGaA, Weinheim, Germany, 2013, Ch. 33, pp. 691–706.

[5] L. Schlapbach, A. Zütel, Hydrogen-storage materials for mobile applications, Nature 114 (2001) 353–358.

[6] M. Appl, Ammonia, in: Ullmann's Encyclopedia of Industrial Chemistry, (Ed.), Wiley-VCH Verlag GmbH & Co. KGaA, Weinheim, Germany, 2006.

[7] A. J. Reiter, S.-C. Kong, Demonstration of compression-ignition engine combustion using ammonia in reducing greenhouse gas emissions, Energy Fuels 22 (2008) 2963–2971.

[8] S. Gill, G. Chatha, A. Tsolakis, S. Golunski, A. York, Assessing the effects of partially decarbonising a diesel engine by co-fuelling with dissociated ammonia, Int. J. Hydrogen Energy 37 (2012) 6074–6083.

[9] J. Hogerwaard, I. Dincer, Comparative efficiency and environmental impact assessments of a hydrogen assisted hybrid locomotive, Int. J. Hydrogen Energy 41 (2016) 6894–6904.

[10] F. R. Westlye, A. Ivarsson, J. Schramm, Experimental investigation of nitrogen based emissions from an ammonia fueled SI-engine, Fuel 111 (2013) 239–247.

[11] S. Frigo, R. Gentili, Analysis of the behaviour of a 4-stroke SI engine fuelled with ammonia and hydrogen, Int. J. Hydrogen Energy 38 (2013) 1607–1615.

[12] K. Ryu, G. E. Zacharakis-Jutz, S.-C. Kong, Effects of gaseous ammonia direct injection on performance characteristics of a spark-ignition engine, Appl. Energy 116 (2014) 206–215.

[13] H. Newhall, E. Starkman, Theoretical performance of ammonia as a gas turbine fuel, in: National Powerplant and Transportation Meetings, SAE International, USA, Feb. 1, 1966.

[14] O. Kurata, N. Iki, T. Matsunuma, T. Inoue, T. Tsujimura, H. Furutani, H. Kobayashi, A. Hayakawa, Performances and emission characteristics of NH_3-air and NH_3-CH_4-air combustion gas-turbine power generations, Proc. Combust. Inst. 36 (2017) 3351–3359.

[15] A. Valera-Medina, R. Marsh, J. Runyon, D. Pugh, P. Beasley, T. Hughes, P. Bowen, Ammonia–methane combustion in tangential swirl burners for gas turbine power generation, Appl. Energy 185 (2017) 1362–1371.

[16] E. J. Cairns, E. L. Simsons, A. D. Tevebauch, Ammonia-oxygen fuel cell, Nature 217 (1968) 780–781.

[17] A. Afif, N. Radenahmad, Q. Cheok, S. Shams, J. H. Kim, A. K. Azad, Ammonia-fed fuel cells: A comprehensive review, Renewable Sustainable Energy Rev. 60 (2016) 822–835.

[18] J. R. Bartels, A feasibility study of implementing an ammonia economy, Graduate theses and dissertations, Iowa State University, Ames, Iowa, USA (2008).

[19] ASHRAE, Handbook of refrigeration, American Society of Heating Refrigerating and Air Conditioning Engineers, Atlanta, GA, USA, 2006.

[20] E. Worrell, L. Price, M. Neelis, C. Galitsky, Z. Nan, World best practice energy intensity values for selected industrial sectors, Technical report, Lawrence Berkeley National Laboratory, Berkeley, CA, USA (2008).

[21] A. Patil, L. Laumans, H. Vrijenhoef, Solar to ammonia - via Proton's NFuel units, Procedia Eng. 83 (2014) 322–327.

[22] M. Reese, C. Marquart, M. Malmali, K. Wagner, E. Buchanan, A. McCormick, E. L. Cussler, Performance of a small-scale Haber process, Ind. & Eng. Chem. Res. 55 (2016) 3742–3750.

[23] A. Schulte-Schulze-Berndt, K. Krabiell, Nitrogen generation by pressure swing adsorption based on carbon molecular sieves, Gas Sep. Purif. 7 (1993) 253–257.

[24] R. Bañares-Alcántara, G. Dericks-III, M. Fiaschetti, P. Grünewald, J. M. Lopez, E. Tsang, A. Yang, L. Ye, S. Zhao, Analysis of islanded NH_3-based energy storage systems, Technical report, University of Oxford, Oxford, UK (2015).

[25] E. R. Morgan, Techno-economic feasibility study of ammonia plants powered by offshore wind, Dissertations, University of Massachusetts, Amherst, Massachusetts, USA (2013).

[26] P. H. Pfromm, Towards sustainable agriculture: Fossil-free ammonia, J. Renewable Sustainable Energy 9 (2017) 034702.

[27] Science & Technology Facilities Council, (accessed Jun. 28, 2018). URL https://stfc.ukri.org/news/uk-team-develop-worlds-first-green-energy-storage-demonstrator/

[28] A. Sánchez, M. Martín, Optimal renewable production of ammonia from water and air, J. Cleaner Prod. 178 (2018) 325–342.

[29] G. Cinti, D. Frattini, E. Jannelli, U. Desideri, G. Bidini, Coupling Solid Oxide Electrolyser (SOE) and ammonia production plant, Appl. Energy 192 (2017) 466–476.

[30] M. Malmali, M. Reese, A. V. McCormick, E. L. Cussler, Converting wind energy to ammonia at lower pressure, ACS Sustainable Chem. Eng. 6 (2018) 827–834.

[31] G. Wang, A. Mitsos, W. Marquardt, Conceptual design of ammonia-based energy storage system: System design and time-invariant performance, AIChE J. 63 (2017) 1620–1637.

[32] M. Penkuhn, G. Tsatsaronis, Comparison of different ammonia synthesis loop configurations with the aid of advanced exergy analysis, Energy 137 (2017) 854–864.

[33] S. Strelzoff, Technology and manufacture of ammonia, John Wiley & Sons Inc, New York, NY, USA, 1981.

[34] M. H. Khademi, R. S. Sabbaghi, Comparison between three types of ammonia synthesis reactor configurations in terms of cooling methods, Chem. Eng. Res. Des. 128 (2017) 306–317.

[35] W. L. Luyben, Design and control of a cooled ammonia reactor, in: G. P. Rangaiah, V. Kariwala (Eds.), Plantwide Control: Recent Developments and Applications, John Wiley & Sons, Ltd, 2012, Ch. 13, pp. 273–292.

[36] M. E. E. Abashar, Application of heat interchange systems to enhance the performance of ammonia reactors, Chem. Eng. J. 78 (2000) 69–79.

[37] W. Nicol, D. Hildebrandt, D. Glasser, Crossing reaction equilibrium in an adiabatic reactor system, Dev. Chem. Eng. Miner. Process. 6 (1998) 41–54.

[38] P. Laššák, J. Labovský, L. Jelemenský, Influence of parameter uncertainty on modeling of industrial ammonia reactor for safety and operability analysis, J. Loss Prev. Process Ind. 23 (2010) 280–288.

[39] C. van Heerden, Autothermic processes, Ind. Eng. Chem. 45 (1953) 1242–1247.

[40] J. C. Morud, S. Skogestad, Analysis of instability in an industrial ammonia reactor, AIChE J. 44 (1998) 888–895.

[41] E. Mancusi, G. Merola, S. Crescitelli, P. L. Maffettone, Multistability and hysteresis in an industrial ammonia reactor, AIChE J. 46 (2000) 824–828.

[42] V. Faraoni, E. Mancusi, L. Russo, G. Continillo, Bifurcation analysis of periodically forced systems via continuation of a discrete map, in: R. Gani, S. B. Jørgensen (Eds.), European Symposium on Computer Aided Process Engineering - 11, Vol. 9 of Comput.-Aided Chem. Eng., Elsevier, 2001, pp. 135–140.

[43] E. Mancusi, P. L. Maffettone, F. Gioia, S. Crescitelli, Nonlinear analysis of heterogeneous model for an industrial ammonia reactor, Chem. Prod. Process Model. 4 (2009) 1–23.

[44] K. Rabchuk, B. Lie, A. Mjaavatten, V. Siepmann, Stability map for ammonia synthesis reactors, in: A. R. Kolai, K. Sørensen, M. P. Nielsen (Eds.), The 55th Conference on Simulation and Modelling (SIMS 55), no. 108 in Modelling, Simulation and Optimization, Linköping University Electronic Press, 2014, pp. 159–166.

[45] M. J. Azarhoosh, F. Farivar, H. Ale Ebrahim, Simulation and optimization of a horizontal ammonia synthesis reactor using genetic algorithm, RSC Adv. 4 (2014) 13419–13429.

[46] F. Farivar, H. A. Ebrahim, Simulation of an axial-radial ammonia synthesis reactor by linking COMSOL-MATLAB software, RSC Adv. 4 (2014) 48293–48298.

[47] J. G. Akpa, N. R. Raphael, Optimization of an ammonia synthesis converter, World Journal of Engineering and Technology 2 (2014) 305–313.

[48] S. Kakaç, H. Liu, A. Pramuanjaroenkij, Heat Exchangers: Selection, Rating, and Thermal Design, 3rd Edition, CRC Press, 2012.

[49] M. Thirumaleshwar, Fundamentals of Heat and Mass Transfer, Always learning, Pearson Education, New Delhi, India, 2009.

[50] D. C. Dyson, J. M. Simon, Kinetic expression with diffusion correction for ammonia synthesis on industrial catalyst, Ind. Eng. Chem. Fundam. 7 (1968) 605–610.

[51] M. Rovaglio, D. Manca, F. Cortese, P. Mussone, Multistability and robust control of the ammonia synthesis loop, in: R. Gani, S. B. Jørgensen (Eds.), European Symposium on Computer Aided Process Engineering - 11, Vol. 9 of Comput.-Aided Chem. Eng., Elsevier, 2001, pp. 723–730.

[52] S. S. E. H. Elnashaie, S. S. Elshishini, Modelling, Simulation and Optimization of Industrial Fixed Bed Catalytic Reactors, Vol. 7, Gordon and Breach Science Publishers, Yverdon, Switzerland, 1993.

[53] M. I. Temkin, N. M. Morozov, E. N. Sheptina, Kinetics of synthesis of ammonia far from equilibrium. ii, Kinet. Catal. 4 (1963) 565–573.

[54] S. Skogestad, Chemical and Energy Process Engineering, Taylor & Francis, Boca Raton, FL, USA, 2008.

[55] L. D. Gaines, Optimal temperatures for ammonia synthesis converters, Ind. Eng. Chem. Process Des. Dev. 16 (1977) 381–389.

[56] S. S. Elnashaie, M. E. Abashar, A. S. Al-Ubaid, Simulation and optimization of an industrial ammonia reactor, Ind. Eng. Chem. Res. 27 (1988) 2015–2022.

[57] S. Elnashaie, F. Alhabdan, A computer software package for the simulation and optimization of an industrial ammonia converter based on a rigorous heterogeneous model, Math. Comput. Model. 12 (1989) 1589–1600.

[58] R. Sinnott, G. Towler, Chemical engineering design: SI edition, Chemical Engineering Series, Elsevier Science, 2009.

[59] L. D. Gaines, Ammonia synthesis loop variables investigated by steady-state simulation, Chem. Eng. Sci. 34 (1979) 37–50.

[60] A. Araújo, S. Skogestad, Control structure design for the ammonia synthesis process, Comput. Chem. Eng. 32 (2008) 2920–2932.

[61] K. V. Reddy, A. Husain, Modeling and simulation of an ammonia synthesis loop, Ind. Eng. Chem. Process Des. Dev. 21 (1982) 359–367.

[62] M. Rovaglio, D. Manca, F. Cortese, A reliable control for the ammonia loop facing limit-cycle and snowball effects, AIChE J. 50 (2004) 1229–1241.

[63] K. Friedrichsen, Energy and economic analysis of the synthesis loop for power-to-ammonia, M.Sc. thesis, Technische Universität Braunschweig, Braunschweig, Germany (2018).

[64] R. N. Brown, Compressors, third edition Edition, Gulf Professional Publishing, Burlington, MA, USA, 2005.

[65] A. K. Sethi, S. P. Sethi, Flexibility in manufacturing: A survey, Int. J. Flex. Manuf. Syst. 2 (4) (1990) 289–328.

[66] T. Seifert, A.-K. Lesniak, S. Sievers, G. Schembecker, C. Bramsiepe, Capacity flexibility of chemical plants, Chem. Eng. Technol. 37 (2014) 332–342.

[67] R. E. Swaney, I. E. Grossmann, An index for operational flexibility in chemical process design. Part I: Formulation and theory, AIChE J. 31 (1985) 621–630.

[68] R. E. Swaney, I. E. Grossmann, An index for operational flexibility in chemical process design. Part II: Computational algorithms, AIChE J. 31 (1985) 631–641.

[69] R. H. Newton, Activity coefficients of gases, Ind. & Eng. Chem. 27 (1935) 302–306.

[70] H. W. Cooper, Fugacities for high P and T, Hydrocarbon Process. 46 (1967) 159–160.

[71] R. S. Herbert, R. W. David, Fugacity coefficients for hydrogen gas between 0 degrees and 1000 degrees C, for pressures to 3000 atm, Am. J. Sci. 262 (1964) 918–929.

[72] L. J. Gillespie, J. A. Beattie, The thermodynamic treatment of chemical equilibria in systems composed of real gases. I. An approximate equation for the mass action function applied to the existing data on the Haber equilibrium, Phys. Rev. 36 (1930) 743–753.

[73] O. A. Hougen, K. M. Watson, Chemical Process Principles, Part III, Wiley, New York, NY, USA, 1962.

[74] M. Shah, Control simulation in ammonia production, Ind. Eng. Chem. 59 (1967) 72–83.

[75] A. T. Mahfouz, S. S. Elshishini, S. S. E. H. Elnashaie, Steady-state modelling and simulation of an industrial ammonia synthesis reactor - I. Reactor modeling., Modelling, Simulation & Control. B 10 (1987) 1–17.

[76] C. G. Alesandrini, S. Lynn, J. M. Prausnitz, Calculation of vapor-liquid equilibria for the system NH_3-N_2-H_2-Ar-CH_4, Ind. Eng. Chem. Process Des. Dev. 11 (1972) 253–259.

[77] B. Poling, J. Prausnitz, J. O'Connell, The Properties of Gases and Liquids, 5th Edition, McGraw Hill professional, McGraw-Hill Education, 2000.

[78] O. Frank, Estimating overall heat transfer coefficients, Chem. Eng. 81 (1974) 126–127.

[79] I. I. Cheema, U. Krewer, Operating envelope of Haber–Bosch process design for power-to-ammonia, RSC Adv. 8 (2018) 34926–34936.

Appendices

A. Supporting equations

A.1. Reactor System

The supporting equations related to heat exchanger and catalyst bed are provided in this section.

A.1.1. Heat exchanger

For shell and tube heat exchanger, effectiveness ε is calculated as:[48]

$$\varepsilon = \frac{2}{1 + C^* + \sqrt{1 + C^{*2}}\dfrac{1 + \exp[-\text{NTU}\sqrt{1 + C^{*2}}]}{1 - \exp[-\text{NTU}\sqrt{1 + C^{*2}}]}} \tag{A.1}$$

where NTU is number of transfer units,

$$\text{NTU} = \frac{U A_{\text{HE}}}{(\dot{m} C_p)_{\text{MIN}}} \tag{A.2}$$

and

$$C^* = \frac{(\dot{m} C_p)_{\text{MIN}}}{(\dot{m} C_p)_{\text{MAX}}} \tag{A.3}$$

A.1.2. Catalyst bed

For solving the equations related to material and heat balance of the catalyst beds, rate of reaction, heat of reaction, constants and parameters are required, and they are calculated as follows:

Mole fraction of components

Mole fractions y_c of components $c \in \{N_2, H_2, NH_3, Ar\}$ are calculated by using a material balance with conversion X_r of reactants, inlet molar fraction y_{in} of the components and stoichiometric coefficients ν of the components and reactants $r \in \{N_2, H_2\}$ as follows:

$$y_{N_2} = \frac{y_{N_2, in} - \frac{\nu_{N_2}}{\nu_r} X_r y_{r_{in}}}{1 - \frac{\nu_{NH_3}}{\nu_r} X_r y_{r_{in}}} \tag{A.4}$$

$$y_{H_2} = \frac{y_{H_2, in} - \frac{\nu_{H_2}}{\nu_r} X_r y_{r_{in}}}{1 - \frac{\nu_{NH_3}}{\nu_r} X_r y_{r_{in}}} \tag{A.5}$$

$$y_{NH_3} = \frac{y_{NH_3, in} + \frac{\nu_{NH_3}}{\nu_r} X_r y_{r_{in}}}{1 - \frac{\nu_{NH_3}}{\nu_r} X_r y_{r_{in}}} \tag{A.6}$$

$$y_{Ar} = \frac{y_{Ar_{in}}}{1 - \frac{\nu_{NH_3}}{\nu_r} X_r y_{r_{in}}} \tag{A.7}$$

Activity

Activity of component a_c is defined as ratio between the fugacity of components $c \in \{N_2, H_2, NH_3\}$ at particular arbitrarily chosen state f_c to the fugacity of pure component f_c^* at pressure 1 bar and temperature equal to system temperature and is calculated as:

$$a_c = \frac{f_c}{f_c^*} = y_c f_c^o = y_c \phi_c P \tag{A.8}$$

where f_c^o is the fugacity at temperature T and pressure P of the system, y_c molar fraction of components and ϕ_c is the fugacity coefficient of components. For nitrogen[69,70], hydrogen[70,71] and ammonia[69,70], the respective fugacity coefficients ϕ_c are as follows:

$$\phi_{N_2} = 0.93431737 + 0.3101804 \times 10^{-3} \, T + 0.295895 \times 10^{-3} \, P$$
$$-0.270729 \times 10^{-6} \, T^2 + 0.4775207 \times 10^{-6} \, P^2 \tag{A.9}$$

$$\phi_{H_2} = \exp\{e^{(-3.8402 \, T^{0.125}+0.541)} P - e^{(-0.1263 \, T^{0.5}-15.980)} P^2$$
$$+ 300 \, [e^{(-0.011901 \, T-5.941)}] \, (e^{-P/300} - 1)\} \tag{A.10}$$

$$\phi_{NH_3} = 0.1438996 + 0.2028538 \times 10^{-2} \, T - 0.4487672 \times 10^{-3} \, P$$
$$-0.1142945 \times 10^{-5} \times T^2 + 0.2761216 \times 10^{-6} P^2 \tag{A.11}$$

Reaction rate constant

The reaction rate constant is expressed by Arrhenius equation as a function of temperature:

$$k = k_0 \, e^{-E_a/RT} \tag{A.12}$$

where k is reaction rate constant, k_0 is frequency factor and E_a activation energy for ammonia decomposition, see table A.1 for values.

Table A.1.: Catalyst properties [50]

α	k_0 / kmol m^{-3}	E_a / kJ kmol^{-1}
0.5	8.8490×10^{14}	1.7056×10^{5}

Equilibrium constant

The equilibrium constant is calculated using the equation developed by Gillespie and Beattie:[72]

$$\log K = -2.691122 \log T - 5.519265 \times 10^{-5} T + 1.848863 \times 10^{-7} T^2$$
$$+ \frac{2001.6}{T} + 2.67899 \tag{A.13}$$

At equilibrium, forward and reverse rates of reaction will be equal, i.e. R_{NH_3} = 0, and at equilibrium for $\alpha = 0.5$, rate of reaction (equation 2.4) reduces to:

$$K^2 = \frac{a^2_{NH_3}}{a_{N_2} a^3_{H_2}} \tag{A.14}$$

where K is the equilibrium constant. For the equilibrium (EQ) line, equations A.13 and A.14 are solved simultaneously for equilibrium conversion X_{EQ} and temperature T_{EQ}.

Specific heat capacity

The specific heat capacities C_{p_c} are expressed as polynomials in T; the pressure dependence for real gases $c \in \{N_2, H_2, Ar\}$ is negligible and calculated as equation A.15 using table A.2.

$$C_{p_c} = 4.184 \left(A_c + B_c\, T + C_c\, T^2 + D_c\, T^3 \right) \tag{A.15}$$

Table A.2.: Coefficients of C_p polynomial for equation A.15 [73]

Component	A_c	$B_c \times 10^{-2}$	$C_c \times 10^{-5}$	$D_c \times 10^{-9}$
N_2	6.903	−0.03753	0.1930	−0.6861
H_2	6.952	−0.04576	0.09563	−0.2079
Ar	4.9675			

For NH_3, pressure dependence is expressed by coefficients of polynomials as linear function of P by Shah.[74] Equation A.16 is taken from data at two pressures, 200 and 600 atm.

$$C_{p_{NH_3}} = 4.184 \{ 6.5846 - 0.61251 \times 10^{-2}\, T + 0.23663 \times 10^{-5}\, T^2$$
$$-1.5981 \times 10^{-9}\, T^3 + [96.1678 - 0.067571\, P + (-0.2225 + 1.6847 \tag{A.16}$$
$$\times 10^{-4}\, P)\, T + (1.289 \times 10^{-4} - 1.0095 \times 10^{-7}\, P)\, T^2] \}$$

The specific heat capacity for gas mixture consisting of components $c \in \{N_2, H_2, Ar, NH_3\}$ in catalyst bed C_{p_b} is:

$$C_{p_b} = \left(\sum_{c=1}^{n} y_{c_b} C_{p_{c_b}} \right) / M_{\text{mix}_b} \tag{A.17}$$

where M_{mix_b} is the average molecular weight of gas mixture.

Heat of reaction

The expression for heat of reaction ΔH is obtained in a similar manner to specific heat of NH_3 (equation A.16) by accommodating pressure dependency of coefficients.[75]

$$\Delta H = 4.184 \left[-(0.54526 + 846.609\, T^{-1} + 459.734 \times 10^6\, T^{-3})P \right.$$
$$\left. -5.34685\, T - 0.2525 \times 10^{-3}\, T^2 + 1.69197 \times 10^{-6}\, T^3 - 9157.09 \right] \tag{A.18}$$

A.2. Ammonia synthesis loop

The material balance for ammonia synthesis loop, shown in figure 2.1 is defined by equations A.19 to A.35, and the percentage yield is calculated by using equation A.36.

Overall material balance

The components and total material balance around the splitter is given as follows:

$$\dot{m}_{c,\textcircled{2}} = \dot{m}_{c,\textcircled{5}} - \dot{m}_{c,\textcircled{6}} \tag{A.19}$$

$$\dot{m}_{\textcircled{2}} = \dot{m}_{\textcircled{5}} - \dot{m}_{\textcircled{6}} \tag{A.20}$$

where components $c \in \{N_2, H_2, NH_3, Ar\}$ concentration in streams $\textcircled{2}$, $\textcircled{5}$ and $\textcircled{6}$ are same, i.e. $x_{c,\textcircled{2}} = x_{c,\textcircled{5}} = x_{c,\textcircled{6}}$ and $y_{c,\textcircled{2}} = y_{c,\textcircled{5}} = y_{c,\textcircled{6}}$, as purging is carried out by splitting stream $\textcircled{5}$ into two streams. Therefore,

$$\dot{m}_{c,\textcircled{2}} = (1 - p)\,\dot{m}_{c,\textcircled{5}} \tag{A.21}$$

where p is defined as ratio between stream $\textcircled{6}$ and stream $\textcircled{5}$:

$$p = \frac{\dot{m}_{\textcircled{6}}}{\dot{m}_{\textcircled{5}}} = \frac{\dot{m}_{c,\textcircled{6}}}{\dot{m}_{c,\textcircled{5}}} \tag{A.22}$$

The material balance around the separator for 100 % pure NH_3 net product ($y_{NH_3,\textcircled{7}} = x_{NH_3,\textcircled{7}} = 1$) and components $c \in \{N_2, H_2, Ar\}$ is given as follows (assumption for chapters 2 and 3 only):

$$\dot{m}_{c,\textcircled{4}} = \dot{m}_{c,\textcircled{5}} \tag{A.23}$$

The total and components material balance around the mixer is stated as follows:

$$\dot{m}_{\textcircled{3}} = \dot{m}_{\textcircled{2}} + \dot{m}_{\textcircled{1}} \tag{A.24}$$

$$\dot{m}_{c,\textcircled{3}} = \dot{m}_{c,\textcircled{2}} + \dot{m}_{c,\textcircled{1}} \tag{A.25}$$

where $\dot{m}$ is mass flow rate for components $c \in \{N_2, H_2, NH_3, Ar\}$ in streams ①, ② and ③. For NH$_3$ free fresh feed ($y_{NH_3,①} = x_{NH_3,①} = 0$), equation A.25 reduces and resulted as follows:

$$\dot{m}_{NH_3,③} = \dot{m}_{NH_3,②} \tag{A.26}$$

Substituting equation A.26 into A.21 results in the following:

$$\dot{m}_{NH_3,③} = (1 - p)\,\dot{m}_{NH_3,⑤} \tag{A.27}$$

In addition, the mass flow rate of components $c \in \{N_2, H_2, NH_3, Ar\}$ in streams ⑤ $\in \{1, 2, 3, 4, 5, 6, 7\}$ of synthesis loop can be expressed as follows:

$$\dot{m}_{c,⑤} = x_{c,⑤}\,\dot{m}_{⑤} \tag{A.28}$$

where x is components mass fraction.

By combining equations A.19, A.22, A.25, A.23 and A.28, we get generalised equation for components $c \in \{N_2, H_2, Ar\}$:

$$\dot{m}_{c,③} - \dot{m}_{c,④} + p\,\dot{m}_{c,④} - x_{c,①}\dot{m}_{①} = 0 \tag{A.29}$$

For fresh feed ① consisting of N$_2$, H$_2$ and Ar, where N$_2$ is supplied along with 2 mol % (0.285 wt %) of Ar, see section 2.2.2. Thus,

$$x_{Ar,①} = 0.0285 x_{N_2,①} \tag{A.30}$$

$$x_{H_2,①} = 1 - x_{N_2,①} - x_{Ar,①} \tag{A.31}$$

By inserting equations A.30 and A.31 into equation A.29 for Ar and H$_2$. The mass balance of components $c \in \{N_2, H_2, Ar\}$ for overall synthesis loop from equation A.29 expressed as follows:

$$\dot{m}_{N_2,③} - \dot{m}_{N_2,④} + p\,\dot{m}_{N_2,④} - x_{N_2,①}\dot{m}_{①} = 0 \tag{A.32}$$

$$\dot{m}_{H_2,\text{③}} - \dot{m}_{H_2,\text{④}} + p\,\dot{m}_{H_2,\text{④}} - (1 - 1.0285 x_{N_2,\text{①}})\dot{m}_{\text{①}} = 0 \qquad (A.33)$$

$$\dot{m}_{Ar,\text{③}} - \dot{m}_{Ar,\text{④}} + p\,\dot{m}_{Ar,\text{④}} - 0.0285 x_{N_2,\text{①}}\dot{m}_{\text{①}} = 0 \qquad (A.34)$$

By solving equations A.32, A.33 and A.34 simultaneously, p, $x_{N_2,\text{①}}$ and $\dot{m}_{\text{①}}$ can be expressed.

The overall total material balance for ammonia synthesis loop is given as follows:

$$\dot{m}_{\text{①}} = \dot{m}_{\text{⑥}} + \dot{m}_{\text{⑦}} \qquad (A.35)$$

Percentage yield

The percentage yield for an overall ammonia synthesis loop is calculated as follows:

$$\text{Percentage yield} = \frac{\text{Mass of actual yield}}{\text{Mass of theoretical yield}} \times 100 \qquad (A.36)$$

where the actual yield is defined as the mass of ammonia in the net product stream and the theoretical yield is defined as the mass of ammonia that is calculated by the chemical equation 1.1. It is important to mention that the theoretical yield is calculated with regard to the limiting reactant. For example, for the minimum NH_3 production and the minimum H_2 intake, the limiting reactant is the H_2, and for the maximum NH_3 production, the limiting reactant is the N_2.

A.3. Separator

Vapour-liquid equilibrium constant $K_{c,S}$ [54] in separators $S \in \{1, 2\}$ is dependent on components $c \in \{N_2, H_2, NH_3, Ar\}$ composition in liquid $y_{c,S_{B_{out}}}$ and vapour $y_{c,S_{T_{out}}}$ phases, which is function of operational pressure P and temperature T. For dilute and supercritical components $c \in \{N_2, H_2, Ar\}$, K_c is expressed by Henry's law as follows:

$$K_{c,S} = \frac{H_{c,S}}{P_S} \qquad (A.37)$$

For NH_3, K_{NH_3} is expressed by Raoult's law:

$$K_{NH_3,S} = \gamma_{NH_3,S} \frac{P^{sat}_{NH_3,S}}{P_S} \qquad (A.38)$$

where H_c is Henry's law coefficients[76] and calculated with equation A.39, $P^{sat}_{NH_3}$ is vapour pressure and calculated with Antoine equation[77]. Furthermore, for equation A.38, $\gamma_{NH_3} = 1$ considered with the assumption of ideal liquid mixture, as liquid phase consist of almost pure ammonia.[54]

$$\ln H_{c,S} = \left(E_c + \frac{F_c}{T_S} + \frac{G_c}{T_S^2} \right) 1.01325 \qquad (A.39)$$

Table A.3.: Constants for calculating Henry's law coefficients (equation A.39)[76]

Components	E_c	$F_c \times 10^4$	$G_c \times 10^6$
N_2	-3.68607	0.596736	-0.692828
H_2	-2.29337	0.529474	-0.521881
Ar	-0.7941	0.447247	-0.509281

$$\log P^{sat}_{NH_3,S} = E_{NH_3} - \frac{F_{NH_3}}{T_S - 273.15 + G_{NH_3}} \qquad (A.40)$$

Table A.4.: Constants for calculating Henry's law coefficients (equation A.39)[77]

Component	E_{NH_3}	F_{NH_3}	G_{NH_3}
NH_3	4.4854	926.132	240.17

The vapour-feed molar fraction β ($0 \leq \dot{n}_{S_{in},L} : \dot{n}_{S_{T_{out}},L} \leq 1$) can be obtained with help of Rachford-Rice flash equation as follows:[54]

$$\sum_c \frac{y_{c,S_{in}}(K_{c,S} - 1)}{1 + \beta(K_{c,S} - 1)} = 0 \qquad (A.41)$$

The dew point can be calculated by fulfilling the underneath condition, with support of P and T adjustments:

$$\sum_c \frac{y_{c,S_{in}}}{K_{c,S}} = 1 \qquad (A.42)$$

B. Additional results

B.1. Operating envelope of Haber-Bosch process design

B.1.1. Reactor system: Heat exchanger

By solving equations (2.1, A.1 to A.3), we obtain heat exchanger effectiveness ε, heat capacity ratio C^*, heat exchanger surface area A_{HE} and number of transfer units (NTU) of the reactor system's heat exchanger shown in figure 2.1 for $U = 50\,\mathrm{W\,m^{-2}\,K^{-1}}$ overall heat transfer coefficient[78], see table B.1.

Table B.1.: Heat exchanger of reactor system

ε / −	C^* / −	NTU / −	A_{HE} / m^2
0.6329	0.5734	1.5984	7.0627

B.1.2. Reactor system: Catalyst beds

Table B.2.: Temperatures and conversions of the reactor system for normal operation and operation at the boundary of the operating envelope

Process Variables	Operating Envelope			Bed-1		Bed-2		Bed-3	
				in	out	in	out	in	out
	Normal Operation (See table 1)		T / K	673.00	773.00	673.00	772.93	673.00	760.01
			X_{H_2} / %	0.00	11.31	7.50	19.19	14.06	24.68
			X_{N_2} / %	0.00	11.31	7.50	19.19	14.06	24.68
			N_2 / mol %	22.71	21.23	21.74	20.10	20.84	19.26
			H_2 / mol %	68.12	63.69	65.24	60.31	62.54	57.79
			NH_3 / mol %	4.17	9.81	7.84	14.11	11.28	17.32
			Ar / mol %	5.00	5.27	5.18	5.48	5.34	5.63
p / bar	194.32	Low	T / K	662.50	740.50	655.63	739.38	657.12	743.42
			X_{H_2} / %	0.00	8.97	5.95	15.88	11.64	22.17
	213.91	High	T / K	679.05	803.00	688.81	792.20	682.48	769.56
			X_{H_2} / %	0.00	13.78	9.13	21.14	15.49	26.09
Y_{Ar} ③ / mol %	0.00	Low	T / K	679.23	785.50	680.55	784.75	680.19	769.86
			X_{H_2} / %	0.00	11.51	7.62	19.26	14.11	24.53
	12.73	High	T / K	654.04	724.85	646.50	723.56	648.50	730.05
			X_{H_2} / %	0.00	8.74	5.79	15.64	11.46	22.21
Y_{NH_3} ③ / mol %	3.39	Low	T / K	678.08	803.00	688.46	790.56	682.17	768.04
			X_{H_2} / %	0.00	13.83	9.16	20.98	15.38	25.78
	4.53	High	T / K	662.96	739.98	655.42	739.57	656.93	744.14
			X_{H_2} / %	0.00	8.89	5.89	15.89	11.65	22.32
$H_2{:}N_2$ ③ / $\frac{\text{mol of } H_2}{\text{mol of } N_2}$	1.18:2.82	Low	T / K	613.89	654.25	606.10	653.13	610.82	666.61
			X_{H_2} / %	0.00	12.41	8.22	23.87	17.34	37.46
			X_{N_2} / %	0.00	1.74	1.15	3.18	2.33	4.74
	3.05:0.95	High	T / K	663.71	743.30	657.23	742.38	658.67	745.33
			X_{H_2} / %	0.00	8.98	5.95	15.84	11.61	21.98
			X_{N_2} / %	0.00	9.61	6.37	17.00	12.46	23.65
$\dot{m}$ ③ / kg h^{-1}	527.78	Low	T / K	675.49	803.00	687.54	787.20	677.98	763.95
			X_{H_2} / %	0.00	14.20	9.41	21.06	15.44	25.97
	707.61	High	T / K	664.02	743.10	657.24	741.97	658.74	745.81
			X_{H_2} / %	0.00	9.08	6.01	16.04	11.75	22.36
T ③ / K	519.41	Low	T / K	661.08	742.07	654.67	741.03	655.66	743.25
			X_{H_2} / %	0.00	9.30	6.16	16.39	12.01	22.72
	536.84	High	T / K	686.02	803.00	696.22	792.28	692.55	772.56
			X_{H_2} / %	0.00	13.05	8.65	19.82	14.52	24.22

B.2. Performance comparison of the autothermal reactor systems

B.2.1. Reactor systems: Heat exchangers

Table B.3.: Data of heat exchangers used in reactor systems

RS	HE	ε / −	C^* / −	NTU / −	A_{HE} / m^2
2Q[79]	HE 1	0.6329	0.5734	1.5984	7.0627
HQ	HE 1	0.5675	0.8758	1.6800	11.2663
	HE 2	0.3680	0.9154	0.5991	4.1103
QH	HE 1	0.4054	0.8182	0.6832	4.1813
	HE 3	0.4821	0.7810	0.9432	5.9077
2H-2	HE 1	0.3398	0.8366	0.5121	4.5437
	HE 2	0.3393	0.9198	0.5244	4.8124
	HE 3	0.5391	0.8888	1.4032	12.5880
2H-3	HE 1	0.3398	0.8366	0.5121	4.5437
	HE 2	0.5007	0.9413	1.1780	10.5640
	HE 3	0.3708	0.8692	0.5967	5.4741

B.2.2. Design and off-design performance

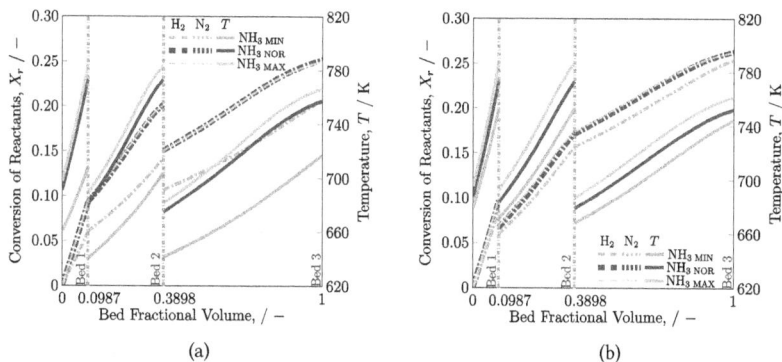

(a)   (b)

Figure B.1.: Reactants conversion and temperature profiles for normal (NOR), minimum (MIN) and maximum (MAX) NH_3 production by varying argon gas composition in (a) reactor system HQ and (b) reactor system QH.

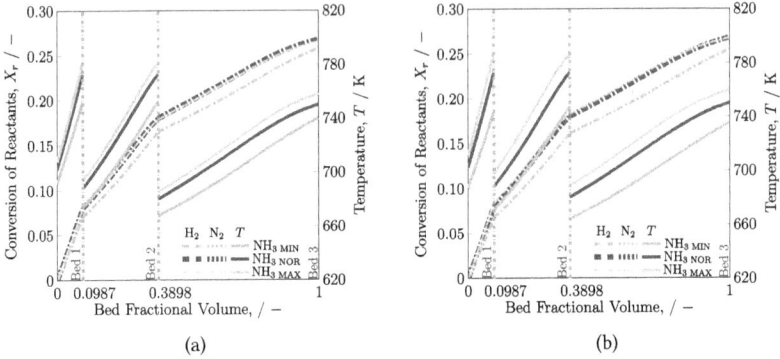

(a)   (b)

Figure B.2.: Reactants conversion and temperature profiles for normal (NOR), minimum (MIN) and maximum (MAX) NH$_3$ production by varying argon gas composition in (a) reactor system 2H-2 and (b) reactor system 2H-3.

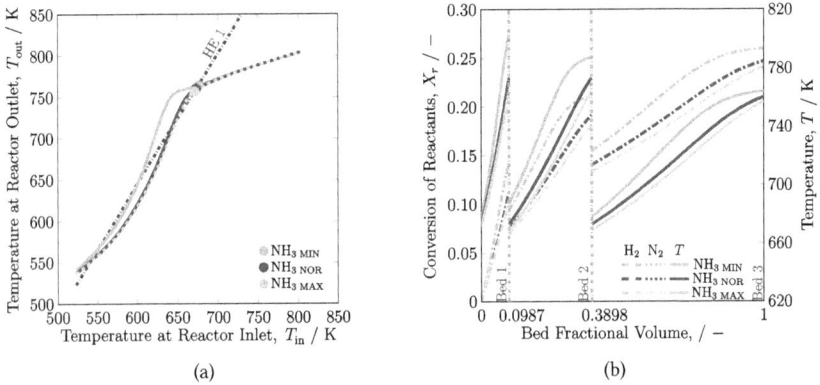

(a)   (b)

Figure B.3.: Direct cooling by quenching reactor system (2Q) for normal (NOR), minimum (MIN) and maximum (MAX) NH$_3$ production by varying process feed flow rate: (a) steady-state characteristics and (b) reactants conversion and temperature profiles.

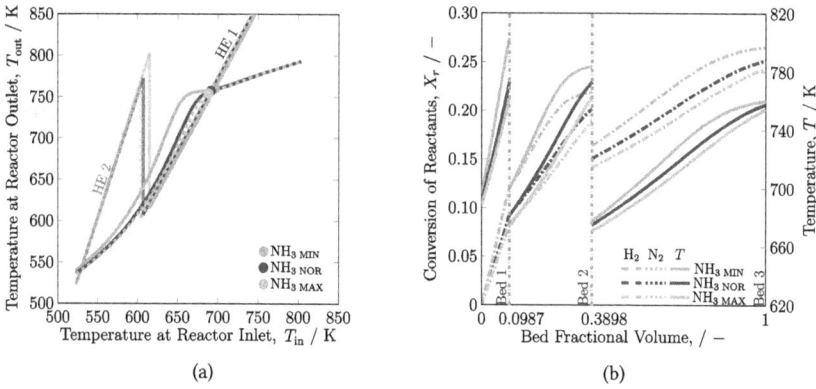

(a)   (b)

Figure B.4.: Combination of indirect and direct cooling reactor system (HQ) for normal (NOR), minimum (MIN) and maximum (MAX) NH$_3$ production by varying process feed flow rate: (a) steady-state characteristics and (b) reactants conversion and temperature profiles.

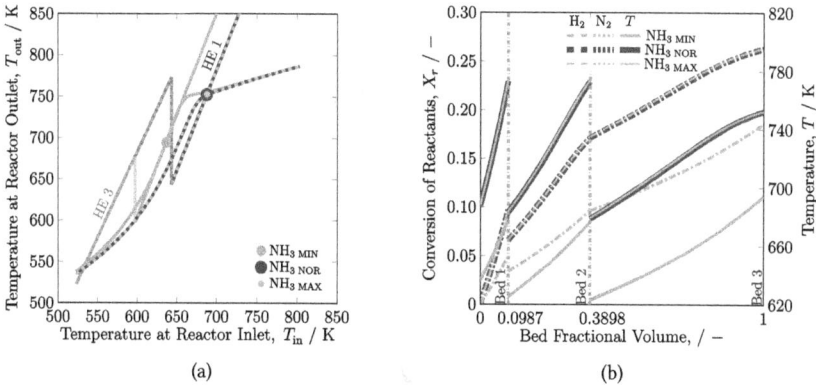

(a)   (b)

Figure B.5.: Combination of direct and indirect cooling reactor system (QH) for normal (NOR), minimum (MIN) and maximum (MAX) NH$_3$ production by varying process feed flow rate: (a) steady-state characteristics and (b) reactants conversion and temperature profiles.

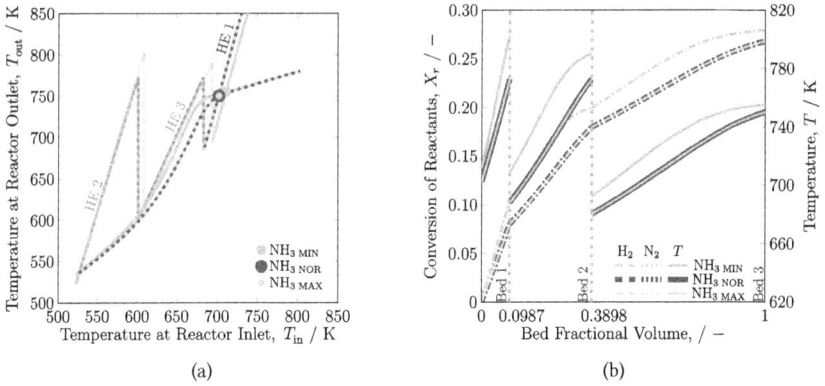

(a)   (b)

Figure B.6.: Indirect cooling by inter-stage heat exchangers (with process feed exchanging heat first in HE 2) reactor system (2H-2) for normal (NOR), minimum (MIN) and maximum (MAX) NH$_3$ production by varying process feed flow rate: (a) steady-state characteristics and (b) reactants conversion and temperature profiles.

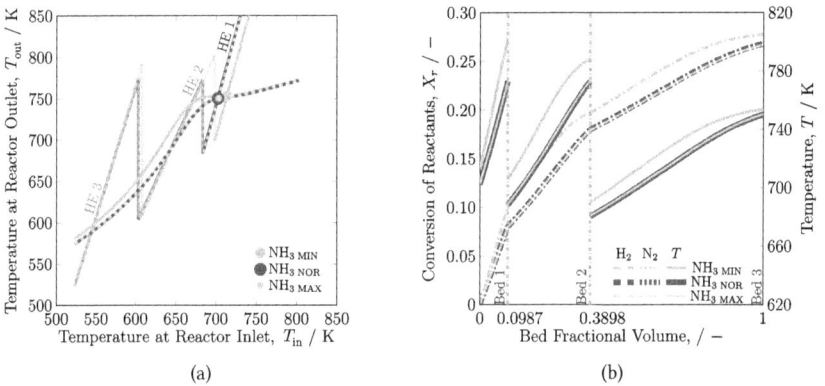

(a)   (b)

Figure B.7.: Indirect cooling by inter-stage heat exchangers (with process feed exchanging heat first in HE 3) reactor system (2H-3) for normal (NOR), minimum (MIN) and maximum (MAX) NH$_3$ production by varying process feed flow rate: (a) steady-state characteristics and (b) reactants conversion and temperature profiles.

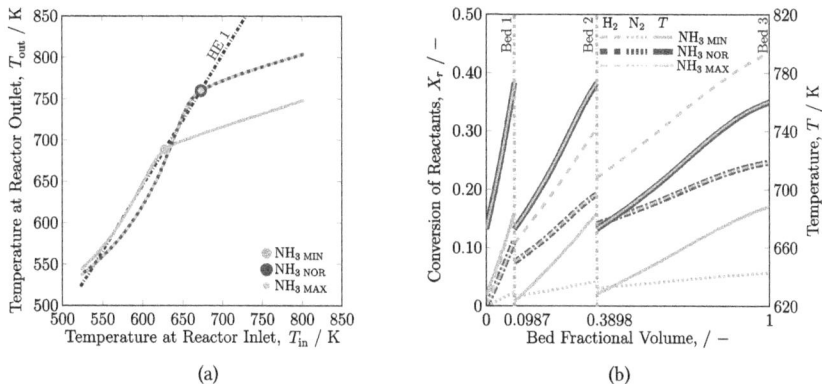

(a)

(b)

Figure B.8.: Direct cooling by quenching reactor system (2Q) for normal(NOR), minimum (MIN) and maximum (MAX) NH_3 production by varying reactants' ratio: (a) steady-state characteristics and (b) reactants conversion and temperature profiles.

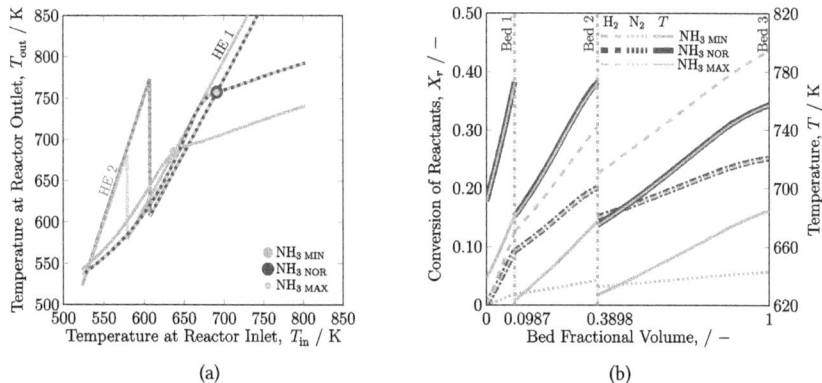

(a)

(b)

Figure B.9.: Combination of indirect and direct cooling reactor system (HQ) for normal (NOR), minimum (MIN) and maximum (MAX) NH_3 production by varying reactants' ratio (a) steady-state characteristics and (b) reactants conversion and temperature profiles.

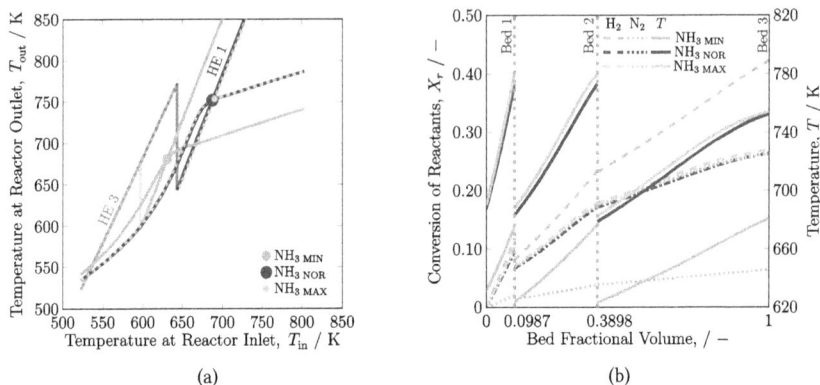

(a)   (b)

Figure B.10.: Combination of direct and indirect cooling reactor system (QH) for normal (NOR), minimum (MIN) and maximum (MAX) NH$_3$ production by varying reactants' ratio: (a) steady-state characteristics and (b) reactants conversion and temperature profiles.

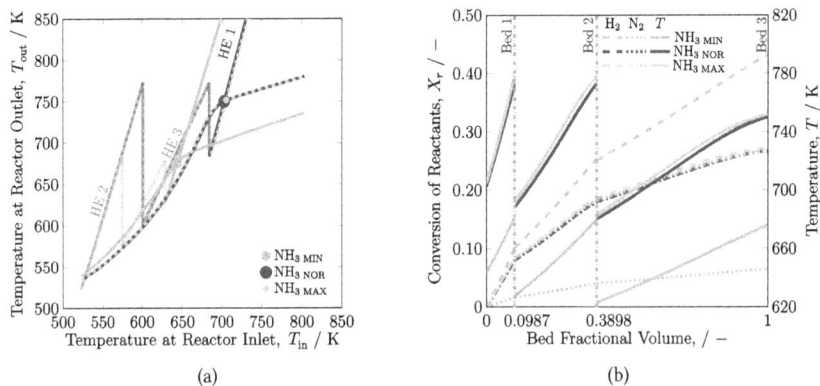

(a)   (b)

Figure B.11.: Indirect cooling by inter-stage heat exchangers (with process feed exchanging heat first in HE 2) reactor system (2H-2) for normal (NOR), minimum (MIN) and maximum (MAX) NH$_3$ production by varying reactants' ratio: (a) steady-state characteristics and (b) reactants conversion and temperature profiles.

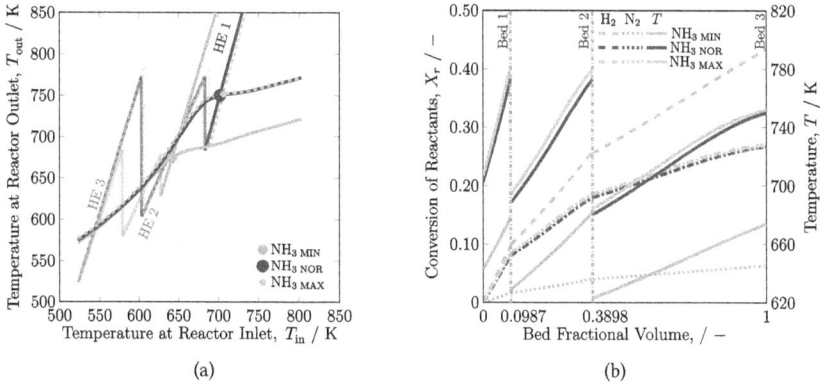

(a)

(b)

Figure B.12.: Indirect cooling by inter-stage heat exchangers (with process feed exchanging heat first in HE 3) reactor system (2H-3) for normal (NOR), minimum (MIN) and maximum (MAX) NH$_3$ production by varying reactants' ratio: (a) steady-state characteristics and (b) reactants conversion and temperature profiles.

B.3. Performance comparison of ammonia synthesis loops

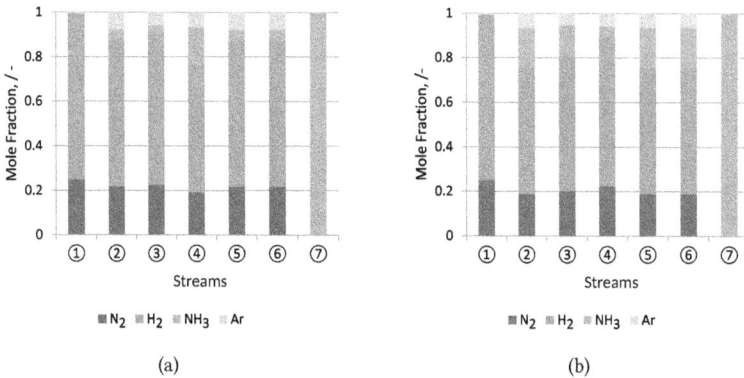

(a)

(b)

Figure B.13.: Mole fraction of components in streams of (a) synthesis loop I and (b) synthesis loop II for normal NH$_3$ production.

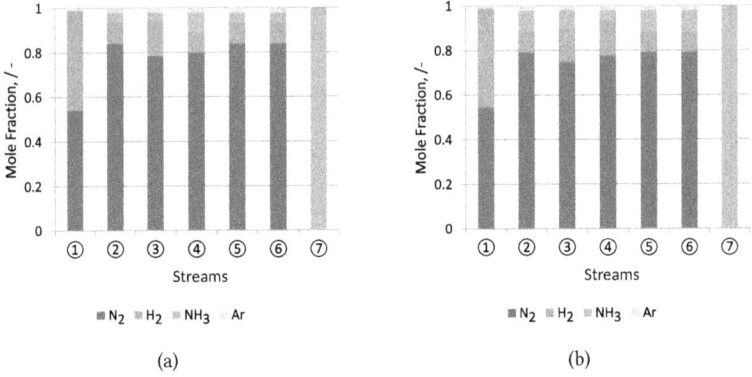

(a)   (b)

Figure B.14.: Mole fraction of components in streams of (a) synthesis loop I and (b) synthesis loop II for minimum NH_3 production.

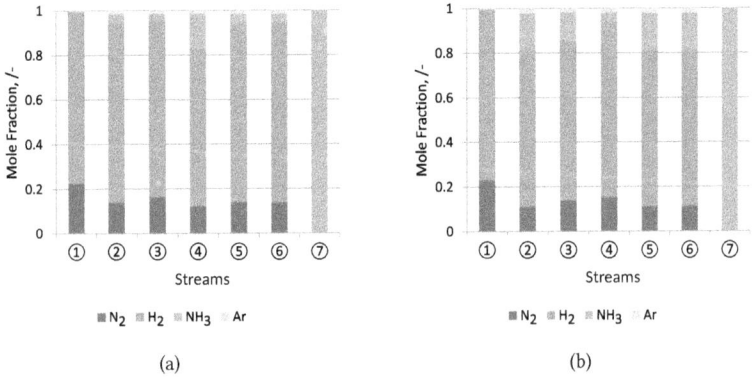

(a)   (b)

Figure B.15.: Mole fraction of components in streams of (a) synthesis loop I and (b) synthesis loop II for maximum NH_3 production.

C. Supporting figures

C.1. Ammonia synthesis loop II

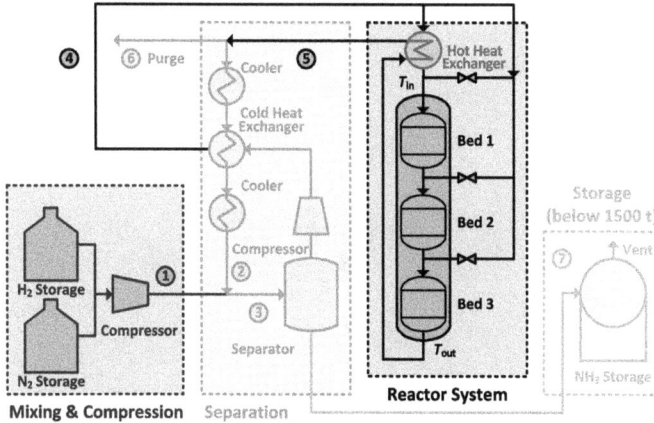

Figure C.1.: Process flow diagram of synthesis loop II, ammonia separation unit before the reactor system.

www.ingramcontent.com/pod-product-compliance
Lightning Source LLC
Chambersburg PA
CBHW060318220326

41598CB00027B/4360